Robert Koch, William Watson Cheyne

Investigations Into the Etiology of Traumatic Infective Diseases

Robert Koch, William Watson Cheyne

Investigations Into the Etiology of Traumatic Infective Diseases

ISBN/EAN: 9783744670869

Printed in Europe, USA, Canada, Australia, Japan

Cover: Foto ©berggeist007 / pixelio.de

More available books at **www.hansebooks.com**

THE NEW SYDENHAM SOCIETY.

INSTITUTED MDCCCLVIII.

VOLUME LXXXVIII.

LONDON:
PRINTED BY WEST, NEWMAN AND CO.,
HATTON GARDEN, E.C.

ETIOLOGY

OF

TRAUMATIC INFECTIVE DISEASES,

BY

DR. ROBERT KOCH,

(WOLLSTEIN.)

TRANSLATED BY

W. WATSON CHEYNE, F.R.C.S.,

ASSISTANT SURGEON TO KING'S COLLEGE HOSPITAL.

CONTENTS.

TRANSLATOR'S PREFACE.

The reader of the following work cannot fail to admit the beauty and importance of the observations which it records, provided he can be satisfied of their authenticity. With regard to the plates which illustrate the text I am able to give the most satisfactory confirmatory evidence. Since the work was published Dr. Koch has succeeded in photographing his microscopic preparations, and he has forwarded a considerable number of the photographs to Professor Lister. These, which have unmistakably been taken from sections of tissue, when examined by a pocket lens or projected on a screen, show plainly that the drawings are faithful representations of what has been seen. I have also lately received from Dr. Koch several of his stained sections, and those which I have been able to examine are equally satisfactory with the photographs.

PREFACE.

The present work pertains to a series of investigations which I have already made regarding the etiology of infective diseases, and which I intend to prosecute more fully hereafter. The object of the enquiry was to determine whether the infective diseases of wounds are of parasitic origin or not. Owing to extraneous circumstances I found it necessary to confine myself solely to experiments on the action of putrid materials on animals. These experiments have led to definite, and, as it seems to me, not unimportant results. Nevertheless, in order to obtain a complete answer to the question it would have been necessary to carry out a further series of similar experiments on animals with materials obtained from persons suffering from or who had died of traumatic infective diseases, and—what indeed seems to me to be the most important—to look for micro-organisms in the human body by the method described in this work.

As, however, I had not the opportunity of completing my investigations in this direction, I have contented myself with producing experimentally in animals morbid processes which resemble the traumatic infective diseases observed in man, and which may serve as illustrations of them.

Since I have succeeded in completing this series of morbid processes, to the extent of furnishing examples illustrative of the most important traumatic infective diseases, *viz.*, of septicæmia, pyæmia, progressive suppuration, gangrene, and erysipelas, I believe that I have solved the problem, so far as is possible by experiments on animals alone; and I therefore think it well to publish the results already obtained.

With respect to the illustrations accompanying this work I must here make a remark. In a former paper* on the examination and photographing of bacteria I expressed the wish that observers would photograph pathogenic bacteria, in order that their representations of them might be as true to nature as possible. I thus felt bound to photograph the bacteria discovered in the animal tissues in traumatic infective diseases, and I have not spared trouble in the attempt. The smallest, and, in fact, the most interesting bacteria, however, can only be made visible in animal tissues by staining them, and by thus gaining the advantage of colour. But in this case the photographer has to deal with the same difficulties as are experienced in photographing coloured objects, *e. g.*, coloured tapestry. These have, as is well known, been overcome by the use of coloured collodion. This led me to use the same method for photographing stained bacteria, and I have, in fact, succeeded, by the use of eosin-collodion, and by shutting off portions of the spectrum by coloured glasses, in obtaining photographs of bacteria which had been stained with blue and red aniline dyes. Nevertheless, from the long exposure required and the unavoidable vibrations of the apparatus, the picture does not have sharpness of outline sufficient to enable it to be of use as a substitute for a drawing, or indeed even as evidence of what one sees. For the present, therefore, I must abstain from publishing photographic representations; but I hope, at a subsequent period, when improved methods allow a shorter exposure, to be able to remedy this defect.

* The paper referred to is published in Cohn's Beiträge zur Biologie d. Pflanzen.

INTRODUCTION.

As at present used, the term "traumatic infective diseases" (Wund infections Krankheiten) indicates a group of affections formerly known as traumatic fever, purulent infection, putrid infection, septicæmia, pyæmia, but which were included at a subsequent period (when the view became generally accepted that these diseases were essentially of the same nature) under the title "pyæmic or septicæmic processes."

Strictly speaking, we should include in this group all those diseases which are the sequel of wounds, even the smallest, as, for example, the pricks of a vaccinating needle, and which have been proved with certainty, by clinical observation or experiment, to be contagious. For example, vaccine infection, anthrax, glanders, hydrophobia, and even syphilis, must be ranked among the traumatic infective diseases. Nevertheless this term does not commonly have such a wide signification, but is limited to those morbid processes specially interesting to the surgeon, which may complicate injuries and operation wounds; in other words, to septicæmia, pyæmia, progressive inflammation and suppuration, and erysipelas. Of late the conviction has more and more gained ground that puerperal fever is also to be regarded as an infective disease, starting from the placental surface or from wounds of the genital passages. Further, many authors include diphtheria also among these diseases, because it at times attacks wounds, and because the possibility of transmitting it by inoculation has been abundantly demonstrated.

In the comments introductory to the experimental part of this work, I shall likewise confine myself to the last-mentioned morbid processes; in the second part, however, I shall deviate from the ordinary limitation of the term "traumatic infective diseases" in so far that I shall also take notice of anthrax on account of its manifold relations to the septicæmia produced experimentally in animals.

The expressions pyæmia and septicæmia are often used with different meanings, and it is therefore necessary to indicate precisely what I shall understand by terms so universally employed.

For a long time pyæmia was distinguished from septicæmia by the occurrence of metastatic deposits in the former and their absence in the latter. But since it has been established that, even in such cases as had been previously described as septicæmia, isolated microscopic metastatic deposits are not unfrequently present, and that the two processes cannot be definitely separated in this way, some authors have preferred to designate as septicæmia the disease brought about by absorption of dissolved putrid poison, and to call all the other morbid processes connected with the development of microscopic organisms, pyæmic processes.

Birch-Hirschfeld,* for example, separates pyæmia from septicæmia in this way. He understands by the term "septicæmia" a disease originating in alterations of the blood, which alterations are a consequence of the absorption of the products of putrefaction. On the other hand, he defines pyæmia as "a general infection, which proceeds from the surfaces of wounds or from the focus of a primary suppurative inflammation, probably evoked by specific organisms and different from the putrid infection." Cohnheim† also identifies

* Lehrbuch der pathologischen Anatomie. Leipzig, Vogel, 876, p. 1224.

† Vorlesungen uber Pathologie. Berlin, 1877, s. 469.

septicæmia with putrid infection, and attributes it to the entrance of a putrid poison in solution into the fluids of the body. Davaine, whose works I shall have to allude to repeatedly, adheres, on the other hand, to the older distinction between pyæmia and septicæmia, and includes under the latter term all those cases in which the post-mortem examination shows no metastatic deposits, although in both cases he considers the co-operation of specific organisms as proven.

The terms pyæmia and septicæmia no longer retain their original signification, for pyæmia does not arise, as was at one time believed, from the entrance of pus into the blood-vessels, and septicæmia is not putrefaction of the living blood. They now remain only as collective names for a number of symptoms which in all probability belong to different diseases. So long, however, as these diseases are not sufficiently separated from each other, it seems best for the present to retain these terms in their ordinary signification, in order to avoid the necessity of constantly adopting new definitions.

For these reasons I shall in what follows include under the term septicæmia all those cases of general traumatic infection in which no metastatic deposits occur, and under pyæmia those in the course of which they may be present.

THE

PRESENT STATE OF KNOWLEDGE

WITH REGARD TO

THE RELATIONS OF MICRO-ORGANISMS TO TRAUMATIC INFECTIVE DISEASES.

I. THE OCCURRENCE OF MICRO-ORGANISMS IN THE HUMAN BODY IN DISEASE.

THE first communication regarding the occurrence of bacteria in the organs of those who have died of traumatic infective diseases was made by Rindfleisch* in the year 1866. In pyæmia, puerperal fever, and like infective diseases, small softened metastatic deposits, of the size of a pin's head, at times occur. These deposits are always found in greater numbers in the substance of the heart, in the muscular tissue of which they have at first the appearance of greyish white spots, these specks becoming, at a later period, cavities filled with a thin fluid pulp. The contents of these cavities consist, as Rindfleisch has shown, not of pus-corpuscles, but solely of "vibriones." These organisms lie at first closely packed between the fasciculi of the muscle; at a later period, however, from disintegration of the muscular fibres, they penetrate into their interior. Rindfleisch was unable to trace the alterations further than to the formation of these small abscess-like softened spots, because the process in question only occurs in those forms of infective diseases which are most severe and rapidly fatal. Rindfleisch has not specially described the organisms termed by him "vibriones," as regards their size or as to whether they were rod-shaped or spherical.

* Lehrbuch der pathologischen gewebelehre. 1 Aufl., s. 204 (4 Aufl., p. 199).

That the development of these miliary purulent deposits, which occur also in other organs in typhus, pyæmia, &c., is produced by parasitic organisms, in other words by bacteria, was shown almost simultaneously by Von Recklinghausen and Waldeyer. Von Recklinghausen* describes the bodies found in the smallest renal veins, in the glomeruli, urinary tubules, and pulmonary alveoli, under the name of micrococci, and states that they may be distinguished from detritus by the uniform size of their granules and by their resistance to the action of glycerine, acetic acid, caustic soda, &c. He likewise calls attention to the brownish colour of the centre of the deposit, as well as to the fact that the uriniferous tubules and the vessels in which the micrococci lie are very much distended at intervals.

Waldeyer confirmed Rindfleisch's statement regarding the occurrence of numerous miliary bacteric deposits in the substance of the heart in pyæmia, and he also found bacteria in small abscess-like spots in the kidneys.

Attention was directed by these investigations to the bacteria, which are present in metastatic deposits in pyæmia, and which up to that time had been overlooked or regarded as unimportant. These statements were confirmed and extended by many similar observations, and it may now be regarded as an established fact that, in most of the metastatic deposits in pyæmia, bacteria in the form of the so-called zooglæa will, on careful examination, be found. Nothing essentially new, with some exceptions to be noticed directly, has, however, been added by later investigations to the original observations of Rindfleisch, Von Recklinghausen, and Waldeyer. It is therefore unnecessary to take special notice of the numerous papers on this subject.

It is worthy of mention that P. Vogt† has seen moving "monads," even during life, in the metastatic deposits of a pyæmic individual. It now naturally occurred to observers to subject the pus of wounds to examination, in order to ascertain if the bacteria, found in the metastatic deposits, accumu-

* Vortrag in der Würzb. physik.-med. Ges. 10 Juni, 1871 (quoted from Birch-Hirschfeld, Med. Jahrbb. Bd. 155, Heft 1).

† Centralblatt für die medicin, Wissenschaft. 1872, No. 44.

lated in the first instance in the pus of the wound, and from thence penetrated into the tissue. Extensive observations of this kind have been made by Birch-Hirschfeld.* He came to the conclusion that the unhealthiness of a wound stood in a direct relation to the number of spherical bacteria in the pus of that wound. The more abundantly these appeared the worse became the state of the wound and the general condition of the patient. The most unfavourable cases were those in which the spherical bacteria had become grouped together in colonies (zooglœa). As the spherical bacteria increased in number their penetration into the pus-corpuscles could also be observed. Birch-Hirschfeld examined at the same time the blood of pyæmic patients, and found that it contained bacteria. He further states that the severity and rapidity of the general infection are in proportion to the number of bacteria which may be detected in the blood.

The channel by which the bacteria gain access to the metastatic deposits would, if these observations were correct, be pretty clearly indicated. The mode by which they pass into the circulation, from the surface of the wound, alone remained unknown. This blank was filled up by the investigations of Klebs. The work of Klebs† deserves notice here, not on this ground alone, but also because his researches furnish very numerous and thorough observations on the bacteria of wound diseases, and further because in them the attempt was made for the first time, by the aid of abundant and excellently used materials for observation, to demonstrate a causal connection between bacteria and traumatic infective diseases. Klebs designates the bacteria found in the pus of a wound as microsporon septicum, starting with the view that spherical and rod-shaped bacteria stand in a genetic relation to each other, and also that the micrococci and bacteria commonly found together in wounds are forms of the same organism. The growth of this microsporon septicum in the form of zooglœa masses firmly attached to the surface of the wound was observed by Klebs on granulations, joint surfaces, and serous membranes. He was also able to trace the penetration

* Untersuchungen über Pyæmie. Leipzig, 1873.

† Beiträge zur patholog. Anatomie der Schusswunden. Leipzig. Vogel, 1872.

of the zooglœa into the interspaces of the cellular tissue. This takes place either with or without the aid of wandering lymph corpuscles. The passage of the microsporon along the lymphatic vessels could not be followed with complete certainty; on the other hand its penetration through the eroded walls of a vein into the circulation was observed in one instance. Further, the elements of the microsporon were found by Klebs in the thrombi which develop behind the valves or veins, and in the metastatic deposits in the lung and liver.

Although the facts which have been hitherto collected with reference to the dependence of pyæmia on the development of bacteria in the body are numerous and important, yet the statements relating to the occurrence of organisms in *septicæmia* are few and doubtful.

Coze and Feltz as well as Hueter* attribute the corrugated form of the red corpuscles frequently seen in septicæmic diseases to the adhesion and penetration of bacteria; an observation which has been much and justly doubted.

The only other statement I have been able to find, as to the presence of bacteria in the blood in septicæmia, is made by Collmann von Schatteburg†. He saw, in one case, rods both in the blood of the body generally and in the vascular loops of the glomeruli.

The observations on erysipelas are much more abundant.

Nepveu‡ found micrococci in the blood of erysipelatous patients, and these were present in greatest number in blood taken from the erysipelatous part.

Wilde§ obtained the same result and he also states that the pus of wounds, from which erysipelatous inflammation starts, contains numerous micrococci. Orth ‖ has also found micrococci in the contents of the bullæ in erysipelas.

Of especial importance is the discovery made by Von Recklinghausen and Lukomsky¶ that the lymphatic vessels

* Compare Birch-Hirschfeld: Lehrbuch der pathologischen Anatomie. Leipzig, 1876, p. 469, and Med. Jahrbb., Bd. 166, p. 184.

† Virchow und Hirsch.: Jahresbericht for 1875. I., page 369.

‡ Virchow und Hirsch: Jahresbericht for 1872. I., page 254.

§ Med. Jahrbb. Bd. 155, Heft 1, page 104.

‖ Archiv f. Experiment. Pathol. u. Pharmakol. I., page 81.

¶ Virchow's Archiv. Bd. 60, page 418.

and canaliculi (saft kanälchen) of the skin at the border of the erysipelatous part are filled with micrococci.

This observation was confirmed by Billroth and Ehrlich,* who likewise found micrococci, not only in the lymphatic, but also in the blood-vessels.

Micrococci have also been seen, by Tillmanns,† in erysipelatous skin, and by Letzerich ‡ in cases of erysipelas attacking vaccination wounds, in the wound itself, in the blood-vessels, muscles, liver, spleen, and kidneys.

With respect to phlegmonous suppurations, the observations have apparently been confined as yet to the contents of the abscesses, while the walls of the latter—that is to say the adjacent tissue which, as will subsequently be shown, is the true seat of the bacteric development—have up to this time received no attention. In the pus from the abscess, just as in the ordinary pus from wounds, bacteria have often and micrococci have almost always been found. A detailed account of the statements on this point is accordingly unnecessary.

Hospital gangrene differs so little from the diphtheria of mucous membranes that the observations made with reference to the latter also hold good for the former.

According to Cohnheim,§ after tracheotomy the disease at times spreads from the mucous membrane to the operation wound. But even without any apparent infection wounds often become diphtheritic, and Cohnheim thinks it probable that hospital gangrene is nothing more or less than a diphtheritic inflammation of the surface of a wound.

Hueter‖ found in the greyish diphtheritic deposits on wounds, and, on more accurate examination, in the neighbouring tissues, as yet apparently quite healthy, the same small round bodies with dark contours which he afterwards saw in the false membrane in diphtheritis of the larynx and pharynx.

* v. Langenbeck's Archiv. Bd. 20, p. 418.

† Deutsche medicin. Wochenschrift, 1878. No. 17, p. 224.

‡ Virchow und Hirsch: Jahresbericht für 1875, p. 69.

§ *L. c.*, p. 482.

‖ Steudener: Volkmann's klinische Vorträge. No. 38, p. 24.

By the researches of Oertel, Nassiloff, Classen, Letzerich, Klebs, and Eberth,[*] it has been placed beyond doubt that in diphtheritic deposits large numbers of micrococci are present. The statements are, however, as yet contradictory with regard to the question whether or not the bacteria penetrate into the tissues.

Oertel[†] found the inflamed mucous membrane crammed with micrococci, and was further able to trace them in the afferent lymphatic vessels of the nearest lymphatic glands, in the glands themselves, as well as in the blood-vessels of the kidney and of other internal organs.

Similar observations have been made by Eberth, Nassiloff, and Letzerich.

The presence of small deposits of bacteria in the cardiac tissue, in the liver, kidneys, and other organs, in cases of diphtheria, has of late been repeatedly demonstrated.[‡]

Attention has also been drawn, by different observers,[§] especially by those who have inoculated diphtheria on the cornea of rabbits, to the brownish colour of the micrococcus masses.

On comparing the behaviour of the bacteria in diphtheria and in pyæmia, one is at once struck by a remarkable correspondence. In both morbid processes the surface of the wound is covered with masses of micrococci which penetrate into the deeper layers of the tissue and into the lymphatic vessels; in both, peculiar miliary bacteric deposits are present in the cardiac tissue, in the liver, and in the kidney; and in both, these bacteric deposits are of a brownish colour. The question is at once forced on the mind, May not the parasitic micro-organisms of pyæmia and of diphtheria be identical?

The same appearances may be seen in puerperal fever. In this disease Waldeyer[‖] has found spherical bacteria in the affected tissues, in the lymphatic vessels, and in the peritoneal exudation; whilst Birch-Hirschfeld[¶] has observed

* *See* Birch-Hirschfeld: Lehrbuch der pathol. Anatomie, p. 799.

† Steudener, *l. c.*, p. 24.

‡ Cohnheim, *l. c.*, page 480.

§ Birch-Hirschfeld, *l. c.*, page 799.

‖ Archiv für Gynäkologie. Bd. II. 1871.

¶ Med. Jahrbb. Bd. 155, p. 105.

micrococcus masses on vaginal ulcers, in the perivaginal cellular tissue, in the blood, in the spleen, and in the liver. The presence of micrococci in the kidneys, lungs, and cardiac muscular tissue was demonstrated by Heiberg and Orth,* and the latter makes mention of the greyish yellow colour of those present in the uriniferous tubules which are affected with nodular dilatations.

As standing probably in close relation to puerperal fever, we must here mention the disease affecting new-born infants, first described by Orth† and called by him *mycosis septica.* In one such case micrococci were found in the blood, in the pleural cavity, and in the urinary bladder.

The so-called mycosis of the navel in new-born infants (nabelmykose) seems likewise to belong to this group of diseases. Weigert‡ describes a case of this kind, and states that the ulcer of the navel was covered with micrococci, and that groups of micrococci were present in the centre of small extravasations of blood in the lungs and kidneys.

Hennig has investigated an analogous case and obtained the same result.

The extremely interesting observations with regard to the occurrence of bacteria in endocarditis seem less easily explicable.

All investigators who have been engaged in seeking bacteria in morbid tissues agree in regarding the undertaking as one of extraordinary, often even of insuperable, difficulty. To make up for the weakness of the anatomical proofs as to the presence of bacteria, pathological experiment has in most cases been resorted to. In order, therefore, to obtain a complete survey of the facts known respecting the relations of bacteria to traumatic infective diseases, it is now necessary to give a short digest of the results of the experimental investigations on this subject.

* *Ibid.* Bd. 166, p. 188.

† Archiv. der Heilkunde, 1872. XIII., p. 265.

‡ Jahresber. der schles. Gesellsch. für vaterl. Kultur. 1875, p. 229.

II. Relations of Micro-organisms to Traumatic Infective Diseases, as established by Experiment.

It is known, as the result of experience, that, when traumatic infective diseases set in, the discharges of the wound and the fluids in the neighbouring tissues take on a putrid character. These alterations in the wound often, indeed, make their appearance before any perceptible manifestation of the general disease, and it was therefore concluded that the putrefaction of the discharges was the cause of the infective disease. Some, however, disputed the accuracy of this conclusion, and maintained that the infective disease was produced by causes acting from within, and that the deterioration of the wound occurred secondarily. To settle this controversy numerous experiments have been set on foot. Experimenters for a long time contented themselves with ascertaining the noxious influence of putrid substances on animals when injected into the blood or into the subcutaneous tissue, and with isolating the poisonous substance contained in these putrid fluids. The question as to whether the disease produced by the injection of the putrid fluid was only a simple poisoning, or whether it in reality possessed the infective qualities of those diseases observed in man, was left untouched by the older and most of the later experimenters. If in an animal, by injection of a putrid fluid, a disease was produced resembling to some extent the human infective disease in symptoms and post-mortem appearances, this circumstance sufficed for their identification, and from such an experiment extensive conclusions as to infective diseases were drawn. But in order that such experiments should prove the infective character of the disease, it must be definitely ascertained, by further transmission from one animal to another, that the disease produced experimentally is in like manner of undoubtedly infective nature.

As we have here to do only with infective diseases, all the investigations which have reference merely to the toxic properties of putrid materials, as well as those in which the possibility of a confusion between toxic action and infection is not excluded, must be left unnoticed.

The first attempt to produce traumatic infective diseases artificially in the lower animals was made by Coze and Feltz.* These investigators injected some grammes of blood, from a patient who had died of putrid poisoning and puerperal fever, into the subcutaneous cellular tissue of rabbits. In consequence of this the animals died with peculiar and characteristic symptoms. A much smaller quantity of the blood of the rabbits thus killed was injected subcutaneously into other rabbits and the same symptoms and fatal termination occurred as took place with the original putrid blood. Coze and Feltz continued this transmission of blood, in gradually diminishing quantities, from the dead animals to others, and they finally succeeded in bringing about infection with an extremely minute amount of blood. This led them to assume that the poison increased in virulence by successive inoculations. In the blood of animals which had died of putrid infection they found bacteria in great numbers, indeed they assert that they have seen at the same time rods, long threads with an oscillating or vermiform motion, and chains of small granules.

The discovery of the increasing virulence of the successively inoculated putrid poison excited the most lively interest.

The experiments of Coze and Feltz were repeated and confirmed by Clementi and Thin, and by Behier and Lionville.† These observers likewise convinced themselves that for the first infection a comparatively large quantity of the infective material, be it blood, peritoneal fluid, or the like, is necessary; while for the later infections an extremely minute quantity is sufficient. They also found numerous bacteria in the blood of the animals killed by the inoculation.

Colin, Vulpian, Raynaud, and others obtained similar results.‡

Davaine,§ however, has studied these conditions more thoroughly than any other observer. He transmitted the

* Virchow und Hirsch: Jahresbericht für 1866. I., p. 195.

† Richter: Die neueren Kenntnisse der Krankmachenden Schmarotzerpilze. separatabdr. aus d. med. Jahrbb.

‡ Med. Jahrbb. Vol. 166, p. 174.

§ *Ibid.*

infection through a series of twenty-five animals in succession, and for the last effective transmission of the putrid infecting material he used only a trillionth part of a drop of blood. Davaine saw in the blood of these animals actively moving bacteria, differing in that respect from those of splenic fever, which are quiescent and were for that reason called by him bacteridia.

Although the most diverse fluids were used by him in these experiments for the first infection, *e. g.*, putrefying blood, blood from pyæmic cases, from puerperal fever, scarlet fever, small-pox, and typhus, yet the effects produced were always similar, and the post-mortem examination showed in all cases bacteria in the blood, and swelling of the spleen, with absence of metastatic deposits. Davaine accordingly calls this disease septicæmia.

Of the other traumatic infective diseases, diphtheria and erysipelas have been artificially produced in animals.

The attempt to transmit erysipelas was first made by Orth.* He injected under the skin of a rabbit the contents of an erysipelatous bulla in which were numerous spherical bacteria. There followed an inflammation entirely analogous to erysipelas in man, and, by the application of the œdematous fluid from the subcutaneous tissue of this animal to a second rabbit, the characteristic progressive inflammation was communicated to the latter. In the œdematous fluid in the subcutaneous cellular tissue, and in the affected parts of the skin of the animals suffering from artificial erysipelas, Orth demonstrated the presence of bacteria in large numbers.

Lukomsky † has also experimented on rabbits with erysipelatous fluid in order to produce an artificial erysipelas. He, however, obtained in the animals experimented on an extensive phlegmonous inflammation of the subcutaneous cellular tissue with implication of the skin. But in his cases also micrococci were present in the canaliculi (saftkanälen) of the areolar tissue and in the lymphatic vessels.

That diphtheria may be communicated to rabbits, and that micrococci appear in the artificial diphtheria similar

* Birch-Hirschfeld: Lehrb. d. patholog. Anat., p. 608.

† *Loc. cit.*

in characters to those found in diphtheria in man, has been demonstrated by Hueter, Tommasi, Œrtel, and Letzerich (*l. c.*)

The investigations on diphtheria led to the introduction of an experiment of extreme importance in the study of pathogenic bacteria, *viz.*, the use of the transparent cornea of the rabbit as the place for inoculation.

Nassiloff* and Eberth† were the first to carry out these corneal inoculations. At first diphtheritic substances were alone used, but it was soon found that the most diverse putrid materials, products of inflammation, and the like, could also be inoculated on the cornea with effect.

Similar experiments have likewise been carried out and in various ways modified by Leber, Stromeyer, Dolschenkow, Orth, and more especially by Frisch.‡

In a successful inoculation of this kind a peculiar stellate patch with conical processes—the so-called "mushroom" appearance (Pilzfigur)—is produced, the centre of which is the point of inoculation. The materials composing this patch are, in diphtheritic inoculations, dense masses of micrococci, which are of a yellow or greenish-brown colour like the micrococci of the diphtheritic and pyæmic deposits in the muscular tissue of the heart and in the kidneys. The "Pilzfigur" is also obtained by inoculation of putrid fluids, and here it consists of rod-shaped bacteria. Frisch further inoculated materials from splenic fever on the cornea of living rabbits, and observed the development of splendid "Pilzfiguren," which consisted only of Bacillus Anthracis.§

In all these experiments the inflammatory appearances in the cornea were exactly in proportion to the development and extent of the bacteria. Eberth found the association of the bacteria with the artificial diphtheria of the cornea so

* Virchow's Archiv. Vol. 50, p. 550.

† Eberth. Bacterische Mykosen. Leipzig, 1872.

‡ Experimentelle Studien über die Verbreitung der Fäulnissorganismen in den Geweben. Erlangen, 1874.

§ Die Milzbrand-bakterien und ihre Vegetation in der lebenden Hornhaut. Vienna, 1876.

‖ *L. c.*, p. 14.

constant that he distinctly says: Without the fungi (*i. e.*, bacteria) no diphtheria.

A special method, and one of much promise for demonstrating the origin of infective diseases from a contagium animatum, was followed by Klebs.* He introduced fluids and other substances, taken from patients who were suffering from or who had died of infective diseases, into thoroughly purified flasks containing cultivating fluids. After development of organisms had occurred in these fluids a small quantity was taken and put into a second vessel containing a similar liquid. With the fluid of the second flask a third was inoculated, and so on through a series sufficiently long to enable him to assume that only an excessively small part of the infective substance originally employed, or indeed none at all, could be present in the last cultivating liquid. The fluids, thus freed from the original infective material, were inoculated on animals. Klebs† has employed this method, which he terms fractional cultivation, more especially with substances from diphtheritic and septic processes, though also with material from various other diseases. The fluids obtained in this manner by fractional cultivation, when inoculated on animals, produced again septicæmia and diphtheria; and Klebs also found, both in the cultivating fluids and in the infected animals, the characteristic micrococci.

In a similar manner Orth‡ has grown in cultivating fluids bacteria from an erysipelatous bulla, and, by the injection of this fluid, has reproduced erysipelas in rabbits.

III. Objections to the conclusiveness of these Facts.

The facts put together in the two foregoing sections are undoubtedly of considerable significance. When taken along with theoretical considerations, and looked at in the light of

* Virchow und Hirsch. Jahresbericht, 1874. Vol. i., p. 359.

† Ueber die Umgestaltung der medicinischen Anschauungen in den letzten drei Jahrzehnten. Rede, gehalten in München bei der 50. Versammlung deutscher Naturforscher. Leipzig. Vogel. 1878.

‡ Birch-Hirschfeld, *l. c.*, p. 608.

the brilliant results of the antiseptic method of treatment, they furnish evidence sufficient to enable many to accept as proved the existence of living infective material, especially in traumatic infective diseases. On the other hand, various objections, of more or less weight, have been urged against this assumption; and a short discussion of these is requisite in order to form a judgment as to the significance of bacteria in traumatic infective diseases.

A considerable number of investigators have advanced the statement that the normal blood and tissues of man and of the lower animals always contain micro-organisms. From this some infer that these organisms are not the cause of the infective disease, but that an abnormal increase in their numbers follows the morbid process, because the fluids of the animal body, when altered by disease, present conditions very favourable for their development. We need not consider these views, which have as yet never been experimentally proved, but which are advanced on theoretical grounds alone. Were it, however, true that bacteria do occur in normal blood, and that the same bacteria—*e. g.*, micrococci—are found, though in unusual numbers, in organs altered by disease, then the possibility of proving that these micrococci were the cause of the disease would be rendered much more difficult, perhaps indeed quite hopeless. We must therefore enquire how far the assertion in question is correct.

Lostorfer, Nedsvetzki, and Béchamp* discovered small moving particles in normal human blood. Lostorfer calls these bodies micrococci, and asserts that he has traced their further development to sarcinæ. Nedsvetzki has given to these particles the name of hæmococci, and he considers them as identical with the bodies described by Béchamp. Béchamp has in numerous papers expressed his views respecting the bodies called by him microzymes. He found these bodies in almost all animal fluids, and, from experiments which he carried on in conjunction with Estor, he concludes that microzymes can, through their physiological activity, bring about the coagulation of the blood, and the lactic, acetic,

* Richter, *l. c.*, p. 12.

and alcoholic fermentations; that they are also active in the transformation of glycogen in the liver, in the development of the embryo in hens' eggs during hatching, and in all possible processes in the animal body. That Béchamp looks on his microzymes as intimately related to bacteria is apparent, because, according to him, the microzymes in the intestine below the iliocæcal valve change normally into bacteria; and at diseased spots of the small intestine—as, *e. g.*, where a tapeworm is attached—bacteria develop immediately from microzymes.

J. Lüders, Bettelheim, Richardson,* and, later, Kolaczek and Letzerich,† also believe that they have seen bacteria in normal human blood.

Tiegel‡ and Billroth§ have attempted in an indirect manner to demonstrate the existence of bacteria in normal animal tissues. They introduced, with certain precautions, fresh portions of muscle, liver, &c., into melted paraffin. The tissues thus enclosed in an air-tight case were, after the lapse of some time, examined to ascertain the presence or absence of bacteria. Numerous bacteria were found, and hence Billroth‖ concludes that in most of the tissues of the body, in greater proportion indeed in the blood, spores of bacteria capable of development are present.

Objections have been urged against the experiments of Billroth and Tiegel, to the effect that the enclosure in paraffin does not protect against the entrance of bacteria, because cracks and fissures form in the paraffin on cooling, and even afterwards, as every one must have observed who has embedded objects in paraffin for microscopical examination.

When normal blood was tested by Pasteur,¶ Burdon Sanderson,** and Klebs,†† as to its power of causing develop-

* Virchow and Hirsch. Jahresbericht for 1868. Vol. i., p. 205.
† Med. Jahrbb. Vol. 168, p. 68.
‡ Virchow and Hirsch. Jahresber., 1874. Vol. i., p. 119.
§ Coccobacteria Septica. Berlin, 1874, p. 58.
‖ *L. c.*, pp. 60 and 187.
¶ Virchow und Hirsch. Jahresberich., 1874. Vol. i., p. 119.
** Cohn. Beiträge zur Biologie der Pflanzen. Vol. i., part 2, p. 194.
†† Med. Jahrbb. Vol. 166, p. 196.

ment, by methods excluding all sources of error, negative results were obtained.

Opposed to the alleged direct observations of bacteria in normal blood are the statements of trustworthy microscopists such as Rindfleisch and Riess,* who distinctly assert that the normal blood is free from bacteria, but that on the other hand, as Riess more especially pointed out, it contains small round bodies more or less numerous, which are most probably *débris* arising from the disintegration of white blood corpuscles, and which, on account of their resemblance to micrococci, have been confounded with them.

According to my own experience the examination of blood, with the view of ascertaining the possible presence of bacteria, is excessively difficult, unless one makes use of the aids to be described afterwards, *viz.*, of staining and suitable illumination. Without the assistance derived from these methods it is in most cases impossible to distinguish the bodies so characteristically described by Riess, from true micrococci; and I can therefore easily imagine that, according as one wished to find that bacteria were present or absent, the granular constituents of the blood would be regarded as micrococci, or micrococci when present would be regarded as the remains of disintegrated white corpuscles. I have, however, on many occasions examined normal blood and normal tissues by means which prevent the possibility of overlooking bacteria, or of confounding them with granular masses of equal size; and I have never, in a single instance, found organisms. *I have therefore come to the conclusion that bacteria do not occur in the blood, nor in the tissues of the healthy living body either of man or of the lower animals.*

On the other hand, the following objections which have been raised against the assumption that bacteria are the cause of traumatic infective diseases seem to me to be well founded. In order to establish this assertion (that they are the cause of traumatic infective diseases) it would be absolutely necessary *that the presence of bacteria in these diseases be proved without exception, and further that the conditions as regards their*

* Virchow und Hirsch. Jahresber., 1872. Vol. i., p. 252.

number and distribution be such as to afford a complete explanation of the symptoms. For, if in some cases of a certain form of infective disease bacteria be found, while in others of like nature they are absent, and if further the bacteria present be so few in number that it is impossible that a severe disease or a fatal termination could thereby be produced, then of course nothing remains but to regard the irregular appearance of the bacteria as depending on chance, and their small number as insufficient as an only cause of the disease in question; in other words it is necessary to assume the presence of some other agency. The observations hitherto made with regard to the occurrence of bacteria in traumatic infective diseases do not, however, in reality fulfil the necessary conditions.

On account of the difficulties, before alluded to, attending the demonstration of bacteria in the blood, and more especially in the tissues, many of the above-mentioned statements have been received with considerable suspicion,—whether always with justice must remain uncertain,—for the earlier method of investigation is in most cases a groping in the dark, and its results cannot be otherwise than very doubtful.

But, apart from the uncertain results of much laborious work on the bacteria of traumatic infective diseases, the literature of the subject contains a number of statements as to complete absence of bacteria in undoubted instances of these affections. It would serve no good purpose to enumerate here all these negative statements, as their value is even less than that of the positive. One or two may, however, be mentioned as illustrations.

Birch-Hirschfeld* states that negative results with regard to the occurrence of bacteria, especially in cases of spreading (fulminanter) gangrene and putrid infection, are by no means rare.

Orth,† after mentioning that micrococci are present in the blood, particularly in septic diseases, puerperal fever, and diphtheria, expressly states that they are by no means constantly found.

* *L. c.*, p. 1224.

† Compendium der pathol.-anatomischen Diagnostik. Berlin, 1876, p. 111.

Eberth,* who has convinced himself of the frequent occurrence of bacteria in septicæmia, does not regard that disease as caused by an infection of the blood by bacteria alone, because he has seen the most distinct septicæmia without the presence of bacteria in the blood.

Weigert† speaks thus of the occurrence of bacteria:—"However certainly one may refer the development of some diseases to the action of bacteria, yet, on the other hand, there exist a far greater number of morbid processes which would theoretically be considered as mycotic, but in which a conscientious observer cannot discover any traces of bacteric agency."

I have intentionally taken these extracts from the writings of those authors who have shown, by having sometimes obtained positive results, that they understood how to overcome the great difficulties connected with the discovery of bacteria, and whose statements, therefore, with regard to their frequent absence in cases of traumatic infective diseases, deserve especial consideration.

The second circumstance, which seems to me of essential importance in considering this question, *viz.*, that in almost all cases in which bacteria were found the number was strikingly small, has as yet received too little attention.

We do not yet know with certainty how many bacteria are necessary to produce in man definite symptoms of disease, or what proportion per kilogram of an animal is required to cause a fatal result. Undoubtedly some definite relations, varying at most only within narrow limits in consequence of differences in the affected individuals, must exist between the number of pathogenic bacteria and their effect, *i.e.*, the symptoms of the disease. The only disease which may with certainty be affirmed to be of bacteric origin, *viz.*, anthrax, gives us sufficient ground for this supposition. Small animals die after inoculation with anthracic blood more quickly than do large ones, and in animals of the same genus and of like size the fatal termination occurs later when the fluid used contains but few spores or bacilli capable of development than when it is rich in them. The explanation of these facts

* Med. Jahrbb. Vol. 166, p. 185.

† Berl. Klin. Wochenschr., 1877, No. 18.

can only be this, that to kill, *e. g.*, a sheep, more bacilli are requisite than to kill a mouse; that bacilli or spores being introduced in about equal quantities into both animals at the time of inoculation, the number of bacilli necessary to cause the death of the mouse, being small, is reached more quickly than the larger quantity requisite to kill the sheep; and further, that in animals of the same species the fatal number of bacilli is longer in being attained when the spores introduced are few in number than when they are injected in large amount.

We learn further from the study of anthrax that the number of bacilli present in the blood must be enormously large before death results. In relapsing fever also, the relations of which to the spirochæta discovered by Obermeier are as yet not sufficiently known, but which nevertheless, on account of the absolutely constant occurrence of these organisms during every attack of fever, must be regarded as in all probability a parasitic disease, the same relation exists between the number of bacteria in the blood and the symptoms of the disease. Of course it must not be assumed that all pathogenic bacteria behave alike in this respect, but it may be concluded from the analogy of anthrax and relapsing fever that a considerable number of bacteria is necessary to produce symptoms. But the observations which have been hitherto made with regard to bacteria in traumatic infective diseases do not in most cases show that this requirement is fulfilled. Mention is generally made of large masses of micrococci on the surface of the wound, which collections, however, can only be regarded as of importance in large wounds; while in internal organs only miliary colonies of bacteria have been found, and these often in small numbers. This result bears no relation whatever to the almost incredible numbers of bacilli in the blood in anthrax. It is therefore only observations which show the presence of a large number of bacteria which can be regarded as affording sufficient explanation of the morbid appearances. Statements no doubt exist with regard to the occurrence of micrococci in the blood and tissues, but unfortunately it is precisely these assertions which for the reasons above-mentioned are the least satisfactory.

A third point remains to be urged against the cogency of the facts known as to the occurrence of bacteria. It is this, that morphologically the bacteria found in the most diverse traumatic infective diseases, and also in other infective processes not in any way connected with wounds, are strikingly similar.

It must at once strike anyone who studies the literature on the subject of bacteria that the two best known bacteric diseases, anthrax and relapsing fever, are notable for the well-marked and easily recognisable form of their parasites; but that in almost all other infective diseases, which apparently stand in close relation with micro-organisms, there exists a remarkable agreement as regards the form, size, arrangement, and colour of the bacteria observed. But, for the very reason that in relapsing fever and in anthrax such marked differences in this respect exist, the similarity of the other pathogenic bacteria must awaken distrust as to the accuracy of the observations and of the assumption that diseases which seem to bear so little relation to each other should, nevertheless, be produced by the same organisms.

Such doubts have been often expressed.

Thus, for example, Birch-Hirschfeld* says, "the morphological characters of the bacteria found in pyæmia, diphtheria, small-pox, and cholera are so similar that the idea naturally arises that identical organisms are being dealt with. But, if this were the case, it would follow that no specific significance could be attributed to these forms. They would have to be regarded merely as parasites of the disease and not as its cause."

To the diseases mentioned by Birch-Hirschfeld a number of others may be added in which micrococci, indistinguishable from each other, were found. Such are erysipelas, puerperal fever, mycosis of the navel in newly-born infants (nabelmycose), hospital gangrene, intestinal mycosis, endocarditis (with or without acute articular rheumatism), primary infective periostitis, scarlet fever, rinderpest, and pleuro-pneumonia. It is, however, impossible that all these diseases can be

* *L. c.*, p. 233.

produced by one and the same parasite, and we must therefore assume either that the micrococci are in reality always the same, in which case they would be merely associated, as an accidental complication, with the diseases enumerated, or that the micrococci—though, on account of their small size, very similar, and indeed apparently the same—are nevertheless different in nature, and consequently capable of giving rise to these diverse results.

In order to show that this latter assumption does not lie beyond the range of possibility, Cohn* has called attention to the apparently similar external and microscopical nature of the sweet and bitter almond, while a great difference exists between them in physiological action. And Virchow,† in the same sense, has referred to the fact that one cannot say with regard to the formative cells of the egg and numerous pathological growths what structure wiil be developed from them, although as compared with bacteria these are truly gigantic in size.

The possibility that the micrococci, in spite of their uniform appearance, may be in reality different, and the true contagium vivum of the disease in which they are found, must assuredly be admitted. But as a practical groundwork, more especially with regard to the prophylaxis and treatment of the traumatic infective diseases, the *possibility* of a contagium vivum does not help us much; we require conclusive evidence that this or that micrococcus, definite in nature and always recognisable under varying conditions by certain characteristics, is the only cause of the disease in question. So long as the existence of a contagium vivum is only a matter of possibility, or even of probability, we cannot avoid taking into consideration in all our investigations, the likewise possible presence of other causes of disease, *e.g.*, of the unknown *x* of a lifeless disease ferment never yet demonstrated, of the *y* of the genus epidemicus, and of other unknown quantities. It is, however, apparent that the solution of the problem proposed would thus be rendered in the highest degree difficult, and would be endangered through numerous sources of error—in fact might probably be quite impossible.

* Beiträge zur Biologie der Pflanzen. Vol. I., part 2, p. 185.

† Die Fortschritte der Kriegsheilkunde. Berlin, 1874, p. 33.

If we now look at the facts brought together and the remarks on them, we come to this conclusion, that the frequent discovery of micro-organisms in traumatic infective diseases and the experimental investigations made in connection with them render the parasitic nature of these diseases probable; but that a thoroughly satisfactory proof has not yet been furnished, and can only be so when we have succeeded in finding the parasitic micro-organisms in all cases of the disease in question, when we can further demonstrate their presence in such numbers and distribution that all the symptoms of the disease may thus find their explanation, and finally when we have established the existence, for every individual traumatic infective disease, of a micro-organism with well-marked morphological characters.

Is it then possible to fulfil these conditions in any degree? or have we now, as many microscopists assume, reached the limit of the capabilities of our optical appliances?

This question will, indeed, have often enough occurred to every one who has specially devoted himself to the examination of pathogenic bacteria. It has also occupied my attention greatly, and at once forced itself on me when I commenced these general investigations on bacteria, and saw what great advantages might be obtained by a proper use of microscopic aids in recognising and distinguishing the smallest forms of bacteria with their spores and cilia.

Since that time I have unremittingly attempted to improve the means for the discovery of pathogenic bacteria in animal tissues, because I could not get rid of the idea that the doubtful results of investigations with regard to the parasites of infective diseases might have their foundation in the incompleteness of the methods used.

I shall now, before passing on to the experimental part of my work, describe the method which I have finally adopted as most suitable, and which has led me to positive results.

METHOD OF INVESTIGATION.

Von Recklinghausen's method, in which advantage is taken of the property bacteria have of resisting acids and alkalies, a power not possessed by animal tissues, is still used by most microscopists. If a group of extremely minute particles, characterised by the uniform size of its component granules, does not alter either in acetic acid or in caustic potash or soda, and if there are other grounds for suspecting the presence of bacteria, the problem may then generally be regarded as solved, and the micrococcus found. An error cannot easily occur here, because the appearance of a closely-packed heap of micrococci, the so-called zooglæa, is so characteristic that he who has once had this picture stamped on his mind will recognise it at any time. Incomparably more difficult are the circumstances when bacteria—and this particularly applies to micrococci—are scattered through the tissues either singly or in small loose groups. For now the characteristic appearance of zooglæa no longer aids the observer, and one has solely to depend on the resisting power of the bacteria to alkalies and acids, because there are a great number of minute particles present in tissues altered by disease which may very easily be confounded with organisms. This method, however, of distinguishing by reagents soon shows itself unreliable. Many bacteria, more especially those which are extremely minute, are as readily destroyed or altered by these reagents as are the animal tissues; and, in the latter, indefinite granules often occur which are not removed by acids or alkalies. This method, therefore, cannot do more than demonstrate zooglæa masses.

An attempt has been made to obtain better results by the use of staining fluids, and the one which has been chiefly recommended, simultaneously indeed by several observers, is hæmatoxylin. This is a great step in advance of the first method, especially when it is used in the same way as for staining nuclei. But it is very incomplete, inasmuch as the hæmatoxylin does not stain rod-shaped bacteria at all, and only colours the spherical so slightly as to prevent their certain recognition when isolated. The use of hæmatoxylin is a great advance on the simple examination of the object with reagents, because when stained the bacteric masses are very much more evident than the rest of the tissue, and one is thus less likely to overlook them or to confound them with other objects. This method, which does not of course exclude examination with reagents, also possesses this great advantage, that the stained preparations can be preserved in Canada balsam, and can thus be used at any time for comparison with others.

Staining with the aniline dyes has yielded still better results than staining with hæmatoxylin. So far as I know, aniline staining, as a means of demonstrating bacteria in animal tissues, was first employed by Weigert. His method, by the communication of which he has laid me under the greatest obligation, is as follows:—

The objects for examination are first hardened in alcohol. The sections made from these are allowed to lie for a considerable time in a pretty strong watery solution of methyl-violet. They are then treated with dilute acetic acid, the water removed by alcohol, cleared up in oil of cloves, and mounted in Canada balsam.

Instead of methyl-violet other aniline dyes, *e.g.*, fuchsin, aniline-brown, &c., may be used in the same manner.

This is, of course, only a general outline of the method; for the individual tissues, and more especially the different forms of bacteria show so great a variety of result from such treatment that it would be impossible to lay down rules which would be universal, and which would apply to every case. For many objects fuchsin is best adapted; for others the methyl colours are more suitable. Among these latter there exists such a difference in the staining power that the sections must lie in one solution only a few minutes, in another several hours. One must therefore work with a large number of sections at once, and test them as to the most suitable

* Bericht uber die Sitzungen der Schlesischen Gesellschaft für Vaterl. Cultur. Dec. 10, 1878.

staining material and as to the time required for the staining. The experienced investigator will find out after a few attempts what is the most suitable material. The strength of the acetic acid solution is not of much consequence. The best solution is one containing only a small percentage of the acid, and it is well not to allow it to act too long. The other manipulations, such as the removal of water, clearing up and mounting, are exactly the same as in the preparation of other microscopic specimens. One must avoid leaving the sections too long in alcohol or oil of cloves, otherwise the staining material will be washed out by these fluids.

In preparations which are treated in this way only the nuclei of the cells and the bacteria are seen to be stained. The latter as a rule take up the aniline colour, and, in fact, their staining is so marked that the individual bacteria can be much more distinctly recognised than after the use of hæmatoxylin. It is thus very easy to recognize with certainty isolated large bacteria, *e.g.*, bacillus anthracis, in the most diverse tissues, when the preparations have been treated with aniline dyes. As soon, however, as we have to deal with smaller bacteria the method yields uncertain results, and, finally, with the smallest forms becomes quite useless.

In order now to understand how it is that small objects, notwithstanding intense staining, cannot be distinguished at all in the animal tissues, or only with difficulty, one must clearly comprehend the component parts of the microscopic picture. Let us for the sake of simplicity consider only the case of a section of an animal tissue mounted in Canada balsam in the ordinary manner.

If all the constituents of this tissue were colourless, and had the same refractive power as Canada balsam, nothing whatever would be seen. This is, however, not the case. Fibres, nuclei, and many other portions of tissue, differ from Canada balsam in their refractive power, and thus by diffraction of the rays of light passing through them an image consisting of lines and shadows is developed which may be termed shortly the structure picture.*

Let us now suppose a second case, namely, that portions

* Nægeli und Schwendener: Mikroskop. Leipzig, 1877, p. 220.

of the tissue, *e.g.*, cell nuclei and bacteria, were coloured, then the conditions would appear as follows:—With equal refractive power of tissue and Canada balsam, nuclei and bacteria would be alone visible, and that on account of the staining material with which they are impregnated; we should therefore have a pure colour picture quite different from the structure picture produced by fibres, membranes, &c., but in part coinciding with that, as for example in the case of the nuclei. For the best possible demonstration of bacteria, which are particularly intensely stained by aniline dyes, such a pure colour picture would certainly be the most suitable. The unavoidable structure picture, however, interferes with this.

Large coloured objects, as for example bacillus anthracis, are but little affected in this way as regards distinctness. Only when the section as a whole or when portions of the tissue are very thick (*e.g.*, the intestinal mucous membrane in its whole thickness) may the structure picture become so preponderant, the number of the shadows placed above each other so great, that even the large bacillus anthracis can be no longer easily distinguished. When, however, the bacteria are smaller and thinner, and thus take up less pigment, the bad effects of the structure picture are much more apparent; a broad dark line may then so overshadow some bacteria that their colour picture becomes too weak to make an impression on the eye. In very thin sections, and in those tissues whose structure consists of but few lines and shadows (*e.g.*, subcutaneous cellular tissue, cornea, &c.), very small bacteria may indeed be distinguished with some accuracy. Ultimately, however, a point is reached where the bacteria are so small that the tiny stained granules and threads are hidden and rendered invisible by even the faintest structure shadows. In some particularly favourable places one may indeed suspect the presence of bacteria, but a sure recognition of them and distinction of their form and size is no longer possible.

This difficulty was experienced by me in these investigations. In materials stained in the manner to be subsequently described I easily found large bacteria, and also smaller ones, particularly when they formed accumulations in the glomeruli of the kidney. But now the thought arose, must not bacteria be present also in the spleen and in the capillaries of the lung? For the spleen was swollen and the blood from the left side of the heart which had just passed through the lungs produced on inoculation on another animal the same fatal disease and the same extremely fine granular accumulations of micrococci in the glomeruli as in the first animal. But in spite of the greatest pains the suspected bacteria could not be found. In the septicæmia of mice, which, as I shall afterwards show, is in the highest degree infective, I was quite unable to demonstrate any micro-organisms. I thus obtained the same incomplete results as the investigators who had formerly studied traumatic infective diseases.

At that time, in attempting to photograph bacteria embedded in Canada balsam, my attention was directed to the fact that the microscopic image consists of a structure and of a colour picture, and I found that the structure picture can be markedly increased or weakened by the nature of the illumination. In this there is nothing new. Every microscopist knows what is the effect of the diaphragm placed below a preparation. A narrow diaphragm not only darkens the field of vision, but makes the structure of the object more apparent; a wide one, on the other hand, renders the whole picture clearer, but makes portions of the structure more indistinct. The difference between narrow and wide diaphragms becomes still more strikingly apparent, when, as in photography, not merely a concave mirror, but a lens or condenser, is used for the illumination, and for this reason, that—particularly with a condenser of short focal distance—the cone of rays illuminating the object is capable of much greater variation. When a narrow diaphragm is placed before the condenser the base of this cone becomes so small that the whole cone may almost be regarded as a bundle of parallel rays of light. The larger, however, the opening of the diaphragm is, so much the larger does the radius of the base of the cone become, its length still remaining the same, but far surpassing as regards the ratio of breadth of base to length that obtained by an ordinary concave mirror. If now we examine a microscopic preparation with an illumination in which the cone of rays is at first narrow, but is gradually made broader, though always remaining of the same length, we shall at once see that—as indeed cannot be otherwise according to dioptrical laws—the structure picture resulting from diffraction, which picture was most marked when the narrowest diaphragm was used, becomes less and less apparent. In proportion, however, as the structure picture diminishes, does the colour picture become more intense and sharply defined. Thus a method is indicated by which the effects of the structure picture may be in so far obviated that even the smallest stained bodies which are within the optical power of the instrument become distinctly discernible. That is to say, an illumination cone must be used of so wide a base that the appearances resulting from diffraction may be completely removed. I tried numerous different lenses and condensers without finding one which removed the structure picture sufficiently till I fell in with the illumination apparatus suggested by Abbe and made by Carl Zeiss in Jena, which I found to answer my purpose in every way.

This apparatus consists of a combination of lenses, the focal point of which is only some millimetres distant from the lowest lens of the objective system. When this compound lens is placed in the opening of the stage of the microscope, a little deeper than the level of the stage, the focal point coincides with the object to be examined, and the latter thus obtains the most favourable illumination. The angle of aperture is so large that on escaping from the condenser into water the outermost rays are inclined at an angle of almost 60° with the axis, the whole effective pencil possessing thus an angle of aperture of 120°—a greater angle than is given any other condenser.* The rays of light are conveyed to this system of lenses by a mirror which is only moveable round a fixed point in the axis of the microscope. Between mirror and lens, and near the focal point of the former, is a support for diaphragms, which are moveable both laterally and circularly, so that the direction of the illuminating pencil may be altered in any way desired. By the use of diaphragms with larger or smaller apertures the aperture of the pencil may be modified from the largest to the smallest attainable with the apparatus. By lateral displacement of the diaphragm without movement of the mirror,

* Naegeli und Schwendener. *L. c.*, p. 99.

oblique illumination can be obtained, and by shutting off the centre of the opening in the diaphragm the middle of the pencil can be got rid of.

By means of this apparatus the connection before described as existing between structure and colour picture can be made evident in the most simple and convincing manner. Let us suppose that a section of a tissue containing very few bacteria stained with aniline is to be examined by the aid of Abbe's illumination apparatus. At first a diaphragm with a narrow opening is used.* The illumination of the object is then about the same as in illumination with a concave mirror and medium cylinder diaphragm. The field of vision, therefore, appears pretty dark, the structure of the tissues is distinctly marked, more especially do the nuclei of the cells strike the eye as dark bodies with a but slightly pronounced staining of a dark blue or red colour; as regards the smaller granules one cannot ascertain at all with certainty whether they are stained or not, nor can it be made out whether these particles are bacteria or constituents of the tissue. Now let diaphragms with a constantly increasing size of aperture be used one after another. The picture gradually alters in a most striking manner. The dark outlines of the cells and cell nuclei, and the sharp lines of the elastic fibres, walls of vessels and the like become pale and ill-defined; the shadows of the bodies present above and below the visual level disappear more and more; many of the points and granules previously seen, which might possibly have been considered to be bacteria, disappear completely, while, on the other hand, small objects which formerly appeared black are observed to be coloured, and the colour of the nuclei becomes more distinct. The field of vision at the same time becomes clearer. The more the lines and shadows and all the differences between clear and dark disappear, so much the more sharply and strongly do all coloured objects stand out, and so much the more distinctly can one recognise their outlines and minute differences in tone and strength of colour. Finally, when the last diaphragm has

* I have had a set of diaphragms prepared, the openings of which increase constantly one millimetre, thus furnishing all varieties of illumination.

been removed, all the outlines of mere structure have disappeared, the field of vision is uniformly cleared, and only coloured objects can be seen. The clearer the light which one selects for illumination (the best light is that from white clouds illuminated by the sun), so much the brighter and more sharply-defined do these objects appear. It then becomes easy to distinguish among the stained bodies the bacteria of which nothing could previously be seen, or which appeared as dark indefinite granules, rods, &c. And this is the more easy, as there is almost nothing stained but nuclei and bacteria. The outlines and size of the bacteria can thus be recognised, and by their uniform appearance they can be distinguished with certainty from other stained granular masses—*e.g.*, broken down cell nuclei.

A very simple arrangement may serve to demonstrate the action of Abbe's illumination apparatus. This consists of a small glass vessel filled with Canada balsam, in which small coloured and colourless glass beads are placed. Here, therefore, conditions are present similar to those in a stained preparation mounted in Canada balsam. The coloured beads correspond to the stained nuclei or bacteria, the colourless to the parts of the tissue which are unstained. If one looks through the vessel on to a broad sheet of paper placed immediately below it and brightly illuminated with daylight, the colourless beads cannot be seen, while the coloured, on the other hand, are distinct and sharply defined. If now the paper be placed at a greater distance from the glass—that is to say, if the pencil illuminating the pearls becomes longer and its angle of aperture smaller, its base remaining the same—the same appearance occurs as when with Abbe's illumination apparatus diaphragms with smaller openings are used in succession, the colourless beads begin to be gradually visible, acquire more distinct and darker outlines, while the coloured ones become darker; and, finally, the two kinds can hardly be distinguished, the coloured becoming completely masked by the colourless. Microscopists who examine for the first time preparations highly magnified and illuminated with Abbe's apparatus without a diaphragm are generally struck with the unusual appearance, the field being too bright and confused, although the outlines of the coloured objects are sharply defined. Such observers have been accustomed to the dark field obtained by ordinary illumination with a concave mirror, and they accordingly feel the want of the outline of the tissue structure. For them it is expedient not to dispense entirely with the diaphragm, but to increase the size of the aperture until the stained object under examination appears sufficiently distinct; there will then remain quite enough of the structure picture to enable them to make out the relations of the tissue to the coloured objects.

It is well in general to use, in addition to the examination of the bacteria by means of the pure colour picture, other methods of investigation, such as the observation of the structure of the tissue at the same time, and the examination of the fresh object, with or without the use of alkalies and acids; and I may here expressly mention that I have often made use of these chemical tests as a check, in addition to my chief method of examination.

Although the aniline staining and the use of Abbe's illumination apparatus so markedly facilitate the investigation of pathogenic bacteria, we must not imagine that all difficulties are thus removed and all sources of error shut out. On the contrary, a considerable amount of practice is necessary before one is in a position rightly to utilise these very efficient means. Some of the difficulties which most frequently occur may be shortly alluded to.

As even isolated bacteria do not escape the observing eye, it not unfrequently happens that one meets with organisms which are derived from the fluids used for staining, washing, &c. For even distilled water is almost never free from bacteria. One, however, very soon learns to distinguish these bacteria from others, and to recognise them immediately as accidental impurities.

Further, incipient putrefaction must be suspected whenever isolated bacteria are found in the superficial layers of organs. But the bacteria appearing in putrefaction, at first generally large bacilli,* are so characteristic that they are not easily

* See my paper on "Photographiren der Bakterien." Beiträge zur Biologie der Pflanzen. Vol. 2, part. 3. Photogramm No. 6. Plate xvi.

confounded with the pathogenic bacteria. Nevertheless, it is well to be cautious in drawing conclusions from preparations already containing putrefactive bacteria; indeed it is best not to use such tissues at all. In order to eliminate every risk of confusion with the putrefactive organisms, and to exclude the idea that in arrangement and number the pathogenic bacteria may have undergone alterations after death, I have only used objects for examination which were placed in absolute alcohol immediately after the death of the animal, though in a few cases a delay of some hours has occurred. Consequently I have never found putrefactive bacteria in the preparations obtained in this way. On the other hand, I have seldom failed to find them in preparations obtained from human subjects, although the *post mortem* examination was made ten to twenty hours after death.

I must here draw attention to a remarkable variety of cells which might give rise to confusion with small masses of micrococcus. These are the so-called plasma cells, described and figured by Ehrlich;* flat cells, for the most part situated on the external coats of blood-vessels, and consisting of a round heap of granules grouped around a nucleus. Their behaviour with respect to aniline staining is exactly the opposite of that of all other cells. In the latter only the nucleus is stained; in the plasma cells, on the other hand, only the finely granular plasma is coloured, the nucleus remaining unstained. Now as the granules have exactly the size of many micrococci, the plasma cell presents the appearance of a small micrococcus colony, more especially when the nucleus is indistinct or has disappeared. But the granules are commonly of unequal size. This fact, taken along with the presence of a nucleus and the results of comparison with other similar cells, enables the diagnosis to be easily made. In human tissues these plasma cells are not very abundant, but they occur in great numbers in mice, particularly in the skin of the ear.

If it be wished to exclude entirely all possibility of confusing bacteria with portions of animal tissues, or if it be desired to render the number and distribution of the bacteria in an organ

* Archiv für Microscopische Anatomie. Vol. xiii. 1877, p. 263.

more evident, then the following method may be made use of:—After staining with aniline, the sections are treated with a weak solution of carbonate of potash, instead of with acetic acid. By this means the nuclei and plasma cells—indeed as a rule all animal tissues—lose again the colouring matter, and the bacteria alone remain stained. Large sections, in which the bacteria only are stained by the method just described, form splendid objects for affording a general view.

In microscopical *technique* staining methods play an important part, and many of the most valuable discoveries already made have been obtained by their help. But the full benefit which staining affords in microscopic work can only, as my investigations show, be completely obtained by making use also of a suitable apparatus for illumination.

This has not yet, so far as I am aware, been taken advantage of, and I do not therefore consider it superfluous to recommend my method of illumination for other microscopical investigations in which it is desired to differentiate very small stained elements from others.

With regard to the method of using Abbe's apparatus, I must draw attention to the fact that a sharply-defined picture can only be obtained by the use of such objective systems as have all the zones of the objective aperture properly corrected. The objective systems made by Zeiss are tested by means of Abbe's condenser as to the correctness of the individual zones, especially of the marginal ones. These, and more especially the new oil systems constructed after the designs of Abbe, are therefore thoroughly adapted for the observation of colour pictures. In other systems which I have tried from this point of view, the marginal zones were almost always insufficiently corrected. The only other lenses with which I have obtained well-defined colour pictures were made by Siebert and Kraft.

ARTIFICIAL TRAUMATIC INFECTIVE DISEASES.

I. Septicæmia in Mice.

Mice are especially adapted for experiments on infective diseases, as I previously found in my investigations on anthrax. I therefore attempted to produce artificial traumatic infective diseases in these animals by the use of the method which was followed by Coze, Feltz, Davaine, &c.

Accordingly putrid fluids, *e.g.*, putrefying blood, putrid meat infusion, &c., were injected under the skin of the back in mice. The result of such an injection differs much according to the nature of the putrid fluid, and according to the quantity which is introduced. Blood and meat infusion, which have putrefied for a long time, appear to act less injuriously than fluids which have putrefied for a few days only. Of these latter fluids, as, for instance, of blood which has not putrefied too long, five drops is sufficient to kill a mouse within a short time. In this case marked symptoms may be observed in the animal immediately after the injection. It becomes restless, running about constantly, but showing great weakness and uncertainty in all its movements; it refuses food, the respiration becomes irregular and slow, and death takes place in four to eight hours.

In such a case the greater part of the fluid injected is found in the subcutaneous cellular tissue of the back in much the same condition as before it was injected. It contains bacteria of the most diverse forms, irregularly mixed together, and as numerous as when examined before injection. No inflammation can be observed in the neighbourhood of the place of injection. The internal organs are also unaltered. If blood taken from the right auricle be introduced into another mouse no effect is produced. Bacteria cannot be found in any of the internal organs, nor in the blood of the heart.

An infective disease has therefore not been produced as the

result of the injection. On the other hand, there can be no doubt that the death of the animal was due to the soluble poison, sepsin, which has been shown by the researches of Bergmann, Panum, and various other investigators, to exist in putrid blood. The animal has accordingly died not from an infective disease, but simply from the effects of a chemical poison.

This supposition is confirmed by the fact that when less fluid is introduced into the animal, the symptoms of poisoning which follow are less marked, and are quite absent when one or at most two drops have been injected. After the use of such small quantities of blood, mice often remain permanently without any morbid symptoms. But a third of them, on an average, become ill after the lapse of about twenty-four hours, during which time they have remained apparently healthy. The symptoms which are then present are characteristic and constant, and are in no case preceded by any of the symptoms of poisoning previously described.

Before I describe these symptoms, I must mention that the infection may succeed when even less than one drop of putrid fluid has been used. The less the amount employed in the first instance, the fewer are the animals affected; for example, of twelve animals inoculated in the ordinary manner with one-twentieth to one-tenth of a drop each, only one was successfully infected.

The first symptom in the infected animals is an increased secretion from the conjunctiva. The eye appears dull, and a whitish mucus collects between the lids, and finally completely glues them together. At the same time lassitude sets in, the animal moves little and languidly; as a rule it sits quite still, with its back much bent and its extremities closely drawn up. It then ceases to eat; its respirations become slower, weakness increases more and more, and death comes on almost imperceptibly. Convulsions never precede it (they always do so in anthrax). After death the animal still remains in the sitting posture with its back strongly bent, while, on the other hand, a mouse which has died from anthrax is always found lying on its back or side with its stiffened extremities fully extended. Thus by the position of the body after death, a fatal result produced by the inoculation of putrefying blood is at once distinguished from that occasioned by inoculation with the material of anthrax. The death of mice infected with putrefying blood occurs forty to sixty hours after the inoculation.

On post-mortem examination there is found at the place of injection or inoculation slight œdema of the subcutaneous cellular tissue. This, however, is often absent, and the internal organs, with the exception of considerable swelling of the spleen, appear quite unaltered.

If one now takes a very small quantity (*e.g.*, one-tenth of a drop) of the fluid of the subcutaneous œdema, or of blood from the heart of such an animal, and inoculates another mouse, exactly the same diseased symptoms occur in the latter animal after the same lapse of time and in the same order as in the former, and death takes place in about fifty hours. From this second animal a third may be infected in like manner, and so on through as many successive animals as one pleases. I have performed these experiments on fifty-four mice and have always obtained the same result. Of these, seventeen inoculations were made in succession; of the others the series of successive infections were less extended.

The certainty with which the infective material can be carried from one mouse to another is here even greater than in anthrax. In the latter, in order to obtain constant results the material for inoculation must be taken from the spleen, because the blood of mice affected with anthrax often contains very few bacilli. In the disease of the mouse produced by putrefying blood it is, on the other hand, a matter of indifference from which organ the material used for inoculation is taken, and even the smallest amount will produce an effect with certainty. It is sufficient, in order to bring about the death of the animal within about fifty hours, to pass the point of a scalpel, which has been in contact with the infected blood, over a small wound of the skin. I have often performed the following experiment:—The subcutaneous tissue of a mouse which had died after inoculation on the tail was touched with a knife on the opposite part of the body to that which had been inoculated, *viz.*, on the head, and with this instrument a small scratch was made on the ear of another mouse. The animals thus infected died, without exception, of the same disease.

This disease is therefore undoubtedly an infective disease, which, from the result of the post-mortem examination, must be called septicæmia.

The great virulence which the blood of septicæmic mice possesses leads us to suspect that if this disease be a parasitic affection brought about by bacteria, the parasites must be present in the blood, and that in great numbers. But in my first investigations I entirely failed to discover bacteria in the septicæmic blood. Not till I used Abbe's condenser did I succeed in demonstrating their presence with complete certainty, in spite of their minute size.

I examined the blood by a method which I have described in another place, and which in this case yielded good results* (by drying it on a cover glass and then staining with methyl violet).

The blood of the animals which became ill after *injection* of one to ten drops of putrefying blood was found to contain as a rule different varieties of bacteria in small numbers, micrococci, and large and small bacilli. If, however, the animals died after *inoculation* with putrefying or septicæmic blood, small bacilli alone appeared in the blood. This result was invariable, and the bacilli were always in large numbers. These bacilli (see Plate i., fig. 1), which lie singly or in small groups between the red blood corpuscles, have a length of ·8 to 1 mikrm.† Their thickness, which cannot be measured accurately, but only approximately estimated, is about ·1 to ·2 mikrm. In order to establish a comparison with other known bacteria there are represented in fig. 4 specimens of Bacillus Anthracis, magnified to the same degree, from the blood of a mouse, the blood having been dried on a cover-glass and stained in exactly the same manner as the septicæmic blood (the lines of junction of the individual rods come out rather too strongly in the drawing). One often sees the bacilli in septicæmic blood attached to each

* Cohn's Beiträge zur Biologie der Pflanzen. Bd. 2, Heft 3, p. 402.

† 1 mikrm. = ·001 millimetre.

other in pairs, either in straight lines or forming an obtuse angle. Chains of three or four bacilli also occur, but they are rare. They show at first sight a great resemblance to small needle-like crystals, but that they are undoubtedly vegetable bodies is evident, for when septicæmic blood is placed on a concave slide and kept in an incubation apparatus the bacilli grow in the same manner as the bacilli of anthrax, not forming, however, long threads like the latter, but dense masses which consist of isolated bacilli. In some cases I have also seen spores appear in the bacilli. I could not, from want of time, study further the conditions of life and of vegetation of these septicæmic bacilli. I intend, however, at some future period to investigate these. Without the use of staining materials the bacilli can only with extreme difficulty be recognised in fresh blood, even when one is familiar with their form, and I have not been able to obtain any certain evidence as to whether they move or not. Their relation to the white blood corpuscles is peculiar. They penetrate into these and multiply in their interior. One often finds that there is hardly a single white corpuscle in the interior of which bacilli cannot be seen. Many corpuscles contain isolated bacilli only; others have thick masses in their interior, the nucleus being still recognisable; while in others the nucleus can be no longer distinguished; and finally, the corpuscle may become a cluster of bacilli breaking up at the margin—the origin of which one could not have explained had there been no opportunity of seeing all the intermediate steps between the intact white corpuscle and these masses (Plate i., fig. 2).

Starting from the point of inoculation, one can easily see the path by which the bacilli have penetrated into the body. In the subcutaneous cellular tissue in the neighbourhood of the inoculated spot they are very numerous, and at times accumulated in dense masses, as can be best observed in inoculations on the ear. They are more especially numerous on the surface of the cartilage of the ear, and they are here covered with a layer of lymph corpuscles. The latter are also present along with numerous red blood corpuscles in the loose cellular tissue.

The large number of red blood corpuscles which pass out of the vessels leads to the conclusion that an alteration has taken place in the walls of these vessels, and thus it becomes extremely probable that the bacilli grow into the vessels and enter the circulation through spaces in their walls, which permitted the exit of the much larger red blood corpuscles. I have never found these bacilli in the lymphatic vessels. Even in the greatly enlarged lymphatic glands they can only be found in the capillary blood-vessels which run through these glands, not, however, in the lymph spaces. In the loose cellular tissue they often spread widely, and may reach from the ear to the mediastinum; from the back into the cellular tissue of the pelvis. I have not found them free in the cavities of the body. Their distribution in the blood-vessels can be best observed on the diaphragm, the vessels running on the border of the centrum tendineum being selected for investigation. The larger veins (Plate iii., fig. 8, shows a small section of one) contain considerable numbers of bacilli pretty equally distributed, and also numerous small clusters developed in the white blood corpuscles. The bacilli which are free in the interior of the vessels are almost always arranged with their long axis in the direction of the blood-stream, and it is thus evident that they were placed in this position by the flowing blood, and after its stagnation have neither increased in number nor moved. In the capillaries the bacilli congregate, particularly at the points of division, but I have never yet seen a complete obstruction of the smaller vessels produced in this way. The inner wall of the arteries is often thickly beset with bacteria directed lengthwise.

In exactly the same manner are the bacilli distributed in the rest of the vascular system. In the examination of sections of lung, liver, kidney, and spleen, one meets everywhere with vessels containing free bacilli, and with white blood corpuscles with bacilli in their interior. The bacilli are not specially accumulated in the glomeruli; strangely enough, they are not more numerous in the greatly enlarged spleen than in other organs.

The whole morbid process has thus a great resemblance to

anthrax. In both diseases the infective power of the blood is due to the bacilli present in it: as soon as these disappear the disease can be no longer produced by inoculation with the blood. Both diseases are distinguished by the invariable development of exceedingly numerous bacilli. There can thus be no doubt that the bacilli of the septicæmia described here possess the same significance as the bacilli of splenic fever, namely, that they are to be regarded as the contagium of this disease.

As anthrax can be successfully inoculated on different species of animals, I have also tried to infect other animals with the blood of septicæmic mice. Having at my disposal only rabbits and field-mice, in addition to house-mice, I was compelled to limit my experiments to them. In both the attempt had a negative result. At first the rabbits were merely inoculated; afterwards the whole of the blood of a septicæmic mouse was injected subcutaneously into one animal, and finally, in addition to the blood, the lungs, heart, liver, kidneys, and spleen of a septicæmic mouse were introduced under the skin of a rabbit.

These animals did not exhibit the slightest evidence of disease, either locally or constitutionally.

It seems peculiar that even field-mice, which resemble house-mice in size and which can hardly at the first glance be distinguished from them, should possess an immunity from this septicæmia. These animals, however, are also much less sensitive to anthrax than house-mice. I attribute this result to differences in the blood of these closely allied animals, which strike one at once on investigation of fresh blood. In the blood of the house-mouse crystals seldom form, and when they do they shoot out only at the border of the drop of blood in the shape of small rectangular tablets and needles. The blood of the field-mouse, on the other hand, always undergoes changes very soon after removal from the body, all the red blood corpuscles becoming transformed into large regular hexagonal plates either immediately or after adjacent corpuscles have run together, and thus the drop becomes in a short time transformed into a crystalline pulp. But although one could not inoculate the septicæmia of the house-mice on the two species of animals mentioned, yet it does not at all follow that all other species likewise possess an immunity from this disease. Many animals are in like manner insensible to anthrax, and it would certainly repay one to test as many different animals as possible with regard to their behaviour towards this septicæmia.

II. Progressive Destruction of Tissue (Gangrene) in Mice.

Occurring along with the septicæmic bacillus just described I have sometimes found in mice, after the introduction of putrefying blood, a micrococcus in the neighbourhood of the place of injection. This organism attracted my attention by its rapid increase and by its regular formation of chains. As a rule, when the animal dies of septicæmia after about two days, none of the numerous forms of bacteria which were injected with the putrid blood can be discovered, except the septicæmic bacilli, or it may be a few residual specimens growing with difficulty. It must therefore be supposed that none of the other bacteria injected at the same time find in the body of the living mouse a suitable soil, and that they therefore perish more or less quickly. My attention was thus at once arrested, when in some cases micrococci were found growing in unusual abundance and of constantly characteristic form. They were not present in the blood, and by inoculation with the blood the septicæmic bacilli alone were transmitted. In order to test whether they could be inoculated, it was therefore necessary that the material used should be taken from the neighbourhood of the place of injection. Inoculations carried out in this way were successful in producing both forms of disease and the virulence of the serum from the subcutaneous cellular tissue containing these micrococci was just as marked as that of the septicæmic blood. When the point of a knife which had been well cleaned was merely brought in contact with the subcutaneous tissue at a spot about one centimetre and a half from the place of injection or inoculation, and when with this knife another animal was immediately inoculated, the inoculation was successful on every occasion. Septicæmia was of course always produced at the same time, because the serum used contained also septicæmic bacilli. The influence of these micrococci on animal tissues and their mode of spreading can be best traced on the ear of a mouse; and it is specially instructive to compare an ear on which only septicæmic bacilli have been inoculated with one into which both the bacilli and the chain-like micrococci have been introduced. In the former ear the cellular tissue is full of red blood corpuscles and lymph cells, so that the bacilli can often be recognized only with great difficulty among the numerous cell nuclei. The other ear presents totally different appearances. Spreading out from the place of inoculation one can see extremely delicate and regular micrococcus chains, here pressed together so as to form thick masses, there arranged diffusely, the individual elements of these chains (Plate iii., fig. 6), as can be estimated from measurements of the longer ones having a diameter of ·5 mikrm. These can be traced almost to the base of the ear, and throughout the part occupied by them all the tissues are markedly altered. As far as the micrococci extend, neither red blood corpuscles, nor nuclei of lymph or of connective tissue cells, can be seen. Even the extremely resistent cartilage cells, and the plasma cells so richly present in the mouse's ear and which are likewise characterised by great resisting power, are pale and scarcely recognisable. All the constituents of the tissue look as if they had been treated with caustic potash; they are dead, they have become gangrenous. Under these circumstances the bacteria develop all the more vigorously. The micrococci penetrate in numbers into the damaged blood and lymphatic vessels, and here and there they fill them so completely that the vessels appear as if injected. Among these the septicæmic bacilli, no longer obscured by nuclei, are seen very distinctly in small groups which at times are very dense and remind one of the "Pilz figures" of the inoculated cornea. While the bacilli can be traced up to the root of the ear, and indeed beyond it, have at the same time increased enormously in the blood, and have ultimately caused the death of the animal, the micrococci, on the other hand, and the destructive process associated with them, have only extended during the same time (within about fifty hours) as far as the vicinity of the root of the ear. Their limit is sharply defined, as can be seen very well on a longitudinal section of the ear examined with a low magnifying power (twenty-five diameters). (Plate i., fig. 5.) The upper part (*c*), from the tip to *b*, is gangrenous. The larger dark oval or round spots (*d*) are transverse sections of vessels containing masses of micrococci in their interior. The widely-distributed micrococcus chains cannot of course be recognised with this power. It is only in the lower fourth of the gangrenous region that they occur in denser groups, which can be seen as little dark points. Then all of a sudden at *b* appears a densely agglomerated mass of nuclei, forming as it were a wall against the invasion of the micrococci, and this is the limit up to which these organisms may be found. They do not extend, even in the blood-vessels, beyond this line. This wall of nuclei has no great breadth, and immediately beyond it comes the normal tissue. By the use of high magnifying powers it becomes apparent that the micrococci do not reach quite up to the nuclear layer. On the side directed towards the micrococci the nuclei are undergoing destruction. Numerous fragments of irregular shape, constantly becoming smaller, form the upper limit of the wall of nuclei, and when this region is reached, in examining the preparation, we may be sure that we are in the neighbourhood of these organisms. There almost always remains between the last remnants of the nuclei and the micrococci a line of considerable breadth consisting only of gangrenous tissue, in which neither micrococci nor nuclei can be found. It is seldom that the micrococci extend into the disintegrating nuclear layer.

These appearances lead us to the conclusion that the action of these micrococci in causing the gangrene is somewhat as follows :—Introduced by inoculation into living animal tissues, they multiply, and as a part of their vegetative process they excrete soluble substances which get into the surrounding tissues by diffusion. When greatly concentrated, as in the neighbourhood of the micrococci, this product of the organisms has such a deleterious action on the cells that these perish and finally completely disappear. At a greater distance from the micrococci the poison becomes more diluted and acts less intensely, only producing inflammation and accumulation of lymph corpuscles. Thus it happens that the micrococci are always found in the gangrenous tissue, and that in extending they are preceded by a wall of nuclei which constantly melts down on the side directed towards them, while on the opposite side it is as constantly renewed by lymph corpuscles deposited afresh.

These observations refer to inoculations with fluid containing both micrococci and bacilli, and it might have been supposed that the septicæmic bacilli were necessary forerunners of the micrococci, that they must to a certain extent prepare the way for them. I therefore attempted, by various means, to separate these parasites from each other. Thus, at one time a considerable quantity, at another only a little of the fluid was used for inoculation, or again it was taken at different distances from the point of inoculation, or, lastly, the parts of the body to which it was applied were varied as much as possible. But all this was of no avail. Either pure septicæmia or septicæmia along with progressive gangrene was obtained, never the latter alone. Chance led me to the proper method. A field-mouse—which, as I formerly pointed out, possesses an immunity from septicæmia—was inoculated with septicæmic bacilli and chain-like micrococci. The experiment was made in the expectation that neither parasite would develop. This expectation, however, was not fulfilled, for, though the bacilli as usual underwent no development, the micrococci increased and spread in exactly the same manner as has been described in the case of the house-mouse. Beginning at the place of inoculation on the root of the tail, the gangrene spread onwards along the back, passing deeply among the dorsal muscles, and downwards on both sides to the abdominal wall. The animal died three days after the inoculation. The parts affected with the gangrene were partially denuded of epidermis and hairs, and contained chain-like micrococci in extraordinary numbers. The same micrococci were also found on the surface of the abdominal organs, although there was no visible peritonitis. The blood and the interior of the organs were, on the other hand, quite free from them. From this animal other field-mice, and from these again house-mice in various successive series were subsequently injected, and always with the like result, *viz.*, that only chain-like micrococci and, in their train, progressive gangrene were obtained.

III. Spreading Abscess in Rabbits.

Coze and Feltz, Davaine, and many others have obtained in rabbits, by the injection of putrid blood, an infective septicæmic disease. I have therefore repeated their experiments. I have not, however, succeeded in producing the effects described by Davaine, but I observed—what others who have made similar experiments on rabbits have already noticed—that in these animals the formation of an abscess constantly increasing in extent may occur in the subcutaneous cellular tissue without any general infection taking place. Such animals have at first no symptoms of disease; a flat lentiform hard infiltration at the seat of the injection is all that can be observed. After several days this hardness extends in all directions, chiefly downwards, especially towards the abdomen and anterior extremities. The animal at the same time emaciates and grows feeble, and dies in about twelve to fifteen days after the injection.

The post-mortem examination shows the presence, in the subcutaneous tissue, of extensive flat abscesses with cheesy contents; their walls bulge in various directions, though the whole remains a single cavity. There is also an extreme degree of emaciation, but no alteration in the peritoneum, intestine, kidneys, spleen, liver, heart, or lungs. In the blood the white corpuscles are greatly increased in number, but no bacteria can be found. The cheesy contents consist of a finely granular material, and scattered about in this are nuclei undergoing disintegration, but no bacteria can be definitely made out. Here, then, we have appearances similar to those often found in man, and much used as an argument against the parasitic nature of such morbid processes. I refer to abscesses resulting from phlegmonous inflammation which must be regarded as infective in their origin, but in which no micro-organisms have been found.

When, however, portions of these abscesses are hardened and examined in sections, the surprising result is obtained that, though bacteria are not present in their contents, their walls are everywhere formed by a thin layer of micrococci united together into thick zooglœa masses. These organisms are the smallest pathogenic micrococci which I have as yet observed. In some places I was fortunate enough to find them arranged in rows, and thus was able to measure them; and I ascertained that they were about ·15 mikrm. in diameter (this is of course only an approximate measurement). From the form and character of the zooglœa masses surrounding the abscesses it follows that these masses stand in the most intimate relation to the contents of the abscesses; that, in fact, the contents are constituted by the zooglœa masses and the dead portions of tissue enclosed by them. This process takes place as follows:—The micrococci grow only in masses which, at the periphery of the more or less lentiform abscess, differ in arrangement from that which they assume on its upper, and more especially on its under surface. The margins of the abscess extend into the loose meshes of the subcutaneous cellular tissue, where the micrococci find the least resistance to their extension, and accordingly surround the abscess in thick cloud-like masses (Plate i., fig. 5). The cellular tissue in the immediate vicinity is more or less richly filled with nuclei (*c*), between which one can see small isolated micrococcus colonies (*b*, *c*)—forerunners, in fact, of the main zooglœa masses. The smallest colonies which can be found seem, from their general form and their radiating pointed processes, to be present in the canaliculi of the cellular tissue. I have not been able to demonstrate any connection between these micrococci and the connective tissue corpuscles, such as is observed in the inoculated cornea. On the wall of that part of the abscess which is directed towards the deeper structures, and where the dense fascia opposes the extension of the organisms, they cannot develop so luxuriantly as at the borders of the abscess. On the contrary, the groups are here small and flattened (Plate iii., fig. 7), and only occasionally send out processes into the layers of the cellular tissue beneath, this tissue being in these situations interspersed with nuclei. An appearance which is quite characteristic can be observed when the zooglœa masses are examined more closely. Their outer borders, by which I understand the parts of the zooglœa masses which are directed towards the healthy cellular tissue (fig. 8, *a*), are stained by the aniline fluid of an intense dark colour, and the individual micrococci can be distinctly made out. In the small, and apparently young colonies more especially (fig. 8, *b*, *c*, *d*, and fig. 7, *b*) the micrococci are uniformly coloured. But on passing towards the interior of the abscess the staining of the zooglœa becomes less marked, the individual micrococci can be no longer accurately defined, they become more and more finely granular, and ultimately form an almost homogeneous mass which no longer takes the colouring matter (fig. 8, *g*).* Still nearer the abscess cavity are found pale unrecognisable masses derived from the zooglœa (fig. 7, *d*) intermixed with the detritus of the nuclei (fig. 7, *e*, and fig. 8, *f*); and the cheesy contents of the abscess are composed of these two materials alone—the dead zooglœa and the remnants of the nuclei, the former being present in largest amount. I have called these unstained masses *dead* zooglœa for the following reasons:—In the first place, this explanation suggests itself so naturally, and seems such a necessary deduction from direct observation and from the comparison of the small micrococcus colonies found in process of growth with the large zooglœa masses which have completed their vegetative life, that no special proof is required for it. One might fairly compare this growth of the micrococci on the one side and their death on the other with the vegetation of Sphagnum. Other considerations further indicate that, when bacteria are no longer stained by aniline, it is a certain sign of their death. The form of bacteric vegetation described here deserves the greatest attention, for it is evident how easily in similar cases the narrow line of bacteria might be overlooked, even though the latter were still in full growth and easily recognisable. Similar circumstances also apparently occur in human infective diseases. Thus Klebs† found in endocarditis that the micrococci deposited on the aortic valves were dark-coloured on the surface, while in the deeper parts

* The micrococci in figs. 7 and 8 have been drawn too large in parts, more especially towards the interior of the zooglœa.

† Archiv fur experimentelle Pathol. und Pharmak. Bd. IX., p. 72 (Taf. II., fig. 3).

they became paler, and finally quite disappeared, passing into a homogeneous mass.

In order to ascertain whether the morbid process here designated as progressive abscess formation could be transmitted from one animal to another, rabbits were injected with blood taken from others which had already died of this disease. These injections produced no effect. A small quantity of the cheesy contents of the abscess was now taken, diluted with distilled water, and injected under the skin of a rabbit. There resulted exactly the same abscess-formation in this animal as in the first. The abscesses spread in the same manner as described in the former case, and caused the death of the animal experimented on in a week and a half. From this animal the disease was conveyed to a third, and so on through several in succession.

It was thus demonstrated that the disease is not merely occasioned by the injection of a considerable quantity of putrefying blood, but is of a decidedly infective character. The assumption made above that the micrococci in the cheesy contents of these abscesses are dead, does not appear in keeping with this result of inoculation. This apparent contradiction may, however, I think, be cleared up, for it is very probable that these micrococci, like other bacteria, form resting spores (Dauer-sporen) after the expiration of their vegetative life, and that these bodies, just like the spores of bacillus, are not stained by aniline, and therefore remain invisible in Canada balsam. The infection in the case referred to would be brought about by such spores.

IV. Pyæmia in Rabbits.

Having failed in various attempts to produce a general infection in rabbits by the injection of putrid blood, I tried the effect of other putrid fluids.

A piece of a mouse's skin about a square centimetre in size was macerated for two days in thirty grammes of distilled water, and a syringeful of this fluid was injected subcutaneously into the back of a rabbit. This animal remained for two days free from any noticeable symptoms of disease, then it began to eat less, became gradually weaker, and died one hundred and five hours after the injection. A post-mortem examination was at once made, and there was found a flat, purulent (not cheesy) infiltration in the subcutaneous cellular tissue, extending from the point of injection as far as the hip behind and the linea alba below. In the abdominal wall the yellowish infiltration extended in parts through the abdominal muscles, and even to the peritoneum. The latter was dull and in many places covered with delicate whitish clots. In the peritoneal cavity a small quantity of turbid fluid was found. The intestines were glued together by white fibrinous masses. The liver, stomach, and spleen were covered with thin white layers of fibrin, and the spleen was much enlarged. The liver, after the removal of the deposit on it, presented a greyish mottled appearance, and showed on section grey wedge-shaped patches; its borders were also in parts of a grey colour. In the lungs were found some dark red patches about as large as a pea, devoid of air. As regards the remaining organs no alterations could be detected, not even in the heart.

A syringeful of blood taken from the heart of this animal was now injected under the skin of the back of a second rabbit. The latter died in forty hours. The result of the post-mortem examination was essentially the same. The infiltration in the neighbourhood of the place of injection was, however, more œdematous, and the cellular tissue was besprinkled with small extravasations of blood; the peritonitis was less advanced; on the small and large intestines a few small subserous extravasations of blood were present; and in the lungs and the liver were metastatic deposits similar to those found in the first rabbit.

I had therefore without doubt a general infective disease to deal with. Indeed it was possible that it might be the same affection as had been obtained by Coze and Feltz, and by Davaine, from injections of putrid fluids into rabbits, and which had led them to their observations on the increasing virulence of septicæmic blood when transmitted through a series of animals in succession. I therefore resolved to carry out a series of experiments similar to those already performed by Davaine.

To furnish a general view of the results of these experiments I shall arrange them in a tabular form :—

Rabbits.	Fluid injected.	Quantity of the same.	Death.
I.	Maceration fluid.	10 drops.	In 105 hours.
II.	Blood of I.	10 ,,	In 40 ,,
III.	Blood of II.	3 ,,	In 54 ,,
IV.	Blood of III.	1 ,,	In 92 ,,
V.	Blood of IV.	$\frac{1}{10}$,,	In 125 ,,
VI.	Blood of V.	$\frac{1}{1000}$,,	Remained well.

The blood was diluted in the same manner as had been done by Davaine in his experiments. In order to obtain the one-thousandth of a drop for injection, a drop of blood was mixed with one hundred drops of distilled water; of this mixture one drop was again added to one hundred drops of distilled water, and of the solution so obtained (the blood being now diluted ten thousand times) ten drops were injected. The table comprises only a few experiments, but the relation between the quantity of blood injected and the time which elapsed before death is so constant that it cannot be accidental. There was no indication of increasing virulence of this blood when inoculated into several animals in succession. The less the quantity of blood injected, the longer the time which elapsed before death occurred, and where the quantity was reduced to the one-thousandth of a drop no result followed. I do not say, however, that the infective power of the blood was entirely abolished by reducing the quantity to one-thousandth of a drop, for this experiment was performed on one animal only, and it might very possibly have happened that if several animals had been injected at the same time with the same amount, one or other would have become ill, or even have died. But it follows from the table that when the quantity is small the effect is delayed, and that when very minute the result is uncertain, or even negative. These results can only be explained on the supposition that the blood always contains a like quantity of undissolved infective particles, and that these particles must have increased to a certain number before they can cause death.

For, an infective material in solution, which is active in quantities varying from ten drops to one-tenth of a drop, must also be infective under all circumstances, even when reduced to a thousandth of a drop; but death would take place at a correspondingly later period.

If, however, the infective material be supposed to be insoluble, and if a certain quantity is always necessary to destroy a rabbit, then the explanation of the fact that the result is more and more delayed in proportion as the injected blood is diluted is at once evident. For the more the blood is diluted the fewer bacteria does each drop contain, and if fewer bacteria be introduced into the animal experimented on, longer time must of course elapse before these have attained the number necessary to cause the death of a rabbit than when the quantity at first injected was large. If the blood be yet further diluted, a time will finally come when in the quantity of blood used for injection, say ten drops, there will not with certainty be a single bacterium, or at least a number sufficient for infection. Then the result becomes a matter of uncertainty.

Let us now see how the facts furnished by microscopic investigation coincide with this explanation.

But first I must mention that the post-mortem appearances found in the last three rabbits were, with some unimportant variations, the same as in the first two, *viz.*, local purulent œdematous infiltration of the subcutaneous cellular tissue, metastatic deposits in the lungs and liver, swelling of the spleen, and peritonitis. These appearances harmonize so completely with those commonly designated as pyæmia, that I do not hesitate to use that term for the disease under consideration.

On microscopic examination micrococci are found in great numbers everywhere throughout the body, and more especially in the parts which have undergone alterations visible to the naked eye. These micrococci are for the most part single or in pairs, and their measurement is therefore difficult. Ten measurements of pairs of micrococci differed but little from each other, and gave ·25 mikrm. as the average diameter of a single individual. As regards size, therefore, they stand midway between the chain-like micrococcus of the progressive gangrene of the tissues and the zooglœa-forming micrococcus of the cheesy abscesses of rabbits. Their relation to the blood-vessels can be best seen in the renal capillaries, and I have therefore selected a small vessel from the cortex of the kidney for delineation (Plate v., fig. 9). It is impossible in a drawing to represent their relative size correctly, and it was necessary to draw them here on a somewhat larger scale than the micrococci of figs. 7 and 8. In the interior of the vessel at *c* is a dense deposit of micrococci adherent to the wall, and enclosing in its substance a number of red blood corpuscles. This mass would probably have very soon filled the calibre of the vessel, for fresh blood corpuscles are constantly being deposited upon it, and these become surrounded by delicate offshoots from the mass of micrococci. From this we may conclude, either that the micrococci have of themselves, owing to the nature of their surface, the power of causing the red blood corpuscles, to which they adhere, to stick together, or that these organisms can occasion coagulation of the blood in their vicinity, and thus the formation of thrombi.

The manner in which these micrococci, as it were, spin round the blood corpuscles and enclose them, seems to me to be quite characteristic of this particular form. Such partial or complete thrombus formations occur in the renal vessels in many places, particularly in the glomeruli, where individual capillary loops may be found completely blocked by micrococci. But even in these thick zooglœa-like masses one can still recognise the clear circles due to the enclosed red blood corpuscles. As a rule, however, only small groups of micrococci are met with (as in fig. 9, *b*). They were found arranged in this manner surrounding and glueing together a small number of blood corpuscles, in the capillary vascular system of all the organs examined, as, for example, in the spleen and in the lungs. In the larger vessels also groups of considerable size are formed, and I am disposed to believe that the large metastatic deposits in the liver and in the lungs do not arise by gradual growth of a mass of micrococci, as in fig. 9, but by the arrest of large groups of micrococci and of the clots associated with them formed in the manner described in the circulating blood; in other words, by true embolism. In the metastatic deposits an extensive development of micrococci occurs, and these are not confined to the vessels, but invade the neighbouring tissues. Micrococci in pairs are pretty equally distributed over the surface of the abdominal organs. Masses of micrococci do not form in the peritonea cavity; and the small flakes of pus in the peritoneal fluid and the fibrinous deposits infiltrated with pus cells present on the surface of the abdominal organs, contain micrococci only uniformly distributed, or at most collected into small groups.

In the neighbourhood of the place of injection the subcutaneous cellular tissue is infiltrated with extensive flat collections of pus, which are surrounded by micrococci more or less numerous, but never in the form of zooglœa. They also surround the subcutaneous veins of this part, which are much distended with blood corpuscles, and their presence in the walls of the vessels and their passage through these walls into the interior can be seen in many places. No micrococci were found in the lymphatic vessels or in the neighbouring lymphatic glands, which were, however, greatly swollen.

Comparing the results of the microscopic investigation with the before-mentioned effects of the injection, we find that they are in complete unison.

In the experiments the blood was taken from the heart, and with reference to the cause of its infecting property we have only to consider the state of the blood in the larger vessels. This, as already mentioned, contains numerous micrococci. The first part of the assumption that the infective particles were bacteria is thus proved. If, however, these underwent the same growth in the blood, as the septicæmic and anthracic bacilli, they must become as numerous in the blood as the latter, and the virulence of the blood would thus be much greater than it was in reality found to be. But, as we have seen, the micrococci of pyæmia behave differently in this respect from the organisms of septicæmia and anthrax. As soon as they come in contact with the red blood corpuscles the latter stick together and form larger or smaller clots in the blood. They can thus no longer pass through the minute capillary networks, like the bacilli which move freely among the red corpuscles, but are arrested in the larger or smaller vessels. From the point of infection fresh micrococci will no doubt constantly pass into the blood, and also individual micrococci will become detached from these small thrombi and emboli and mix with the blood stream. Nevertheless, their total number in the circulating blood cannot exceed a certain point, because they are very soon deposited somewhere. Thus we have a simple explanation of the fact that the number of micrococci present in the body of the animal experimented on constantly increases, and, finally, apart from the disturbances of circulation produced by them, become sufficiently numerous to cause the death of the animal. Nevertheless, the quantity present in the cardiac blood continues pretty uniform, and is so small as to have an uncertain action when the thousandth part of a drop is used.

5. Septicæmia in Rabbits.

After injection of putrid infusion of meat into rabbits, I have twice obtained a general infection of another sort in which metastatic deposits do not occur, and which I would therefore describe, in contrast to the foregoing, as septicæmia. This infusion, like the putrid fluids used in the earlier experiments, contained numbers of bacteria of the most various forms. When injected under the skin of the back of a rabbit it produces an extensive putrid suppuration of the subcutaneous cellular tissue, and the animal dies in three days and a half. At the ichorous spot, which must, on account of its size, be looked upon as the immediate cause of death (owing to absorption of poisonous materials in solution), the same variety of bacteric forms was present as in the meat infusion. At the border of this spot the cellular tissue was infiltrated with a slightly turbid watery fluid, which contrasted strikingly with the brownish stinking ichor in the vicinity of the place of injection. In this œdema fluid great numbers of micrococci of considerable size and of an oval form were almost the only organism observed. In the blood also similar micrococci were found, though only in small numbers. Further, in the papilli of the kidney and in the greatly-enlarged spleen some of the small veins were completely blocked for short distances with these oval micrococci.

Two drops of this œdematous fluid were now injected under the skin of the back of a second rabbit. The animal died in twenty-two hours, and here, in the neighbourhood of the place of injection, not a trace of pus could be observed. On the other hand slight œdema, with a streaky whitish appearance of the subcutaneous cellular tissue, extended from the point of injection to the abdomen. In this œdematous cellular tissue lay numerous flat extravasations of blood half a centimetre in breadth, the vessels around these being very greatly distended. The muscles of the thigh and of the abdominal wall were also interspersed with small extravasations. In the heart and lungs no alterations were found. In the peritoneal cavity no fluid was present, the peritoneum being unaltered and the coils of intestine not glued together. But the surface of the intestine, in consequence of a number of small subserous extravasations presented an appearance as if injected here and there with blood. The spleen was also very considerably enlarged.

In this second animal the oval micrococci were alone present in the œdematous cellular tissue, all the other bacteria having disappeared. The number of these organisms was very considerable, many of the small veins being completely filled with them. In the hæmorrhagic spots were small veins, which were here and there distended with micrococci, thus presenting spindle-shaped dilatations, which had at parts burst, the micrococci having thus escaped in large numbers into the surrounding cellular tissue. This appearance could be particularly well seen in the muscles of the thigh.

In the pulmonary capillaries the micrococci were not very numerous; they were scattered through the blood singly or in pairs, and occasionally in small groups. In the kidneys these organisms were present in much larger numbers. The great majority of the glomeruli seemed enlarged, as if swollen; their capillary loops were increased in size and distended with red blood corpuscles. The other glomeruli were smaller than usual, the nuclei of their capillary walls being closer together, so that they presented an appearance as if they had been compressed.

In all the enlarged glomeruli, without exception, more or less extensive deposits of oval micrococci were present. These were arranged in longitudinal series and also side by side, so as to form a single layer, covering the inner wall of the capillaries for short distances, but never embracing the whole circumference. The micrococcus colonies thus presented the appearance of short, slightly-curled, trough-like pellicles. In other places the vascular loops were completely distended, and there were also present all transitions from these dense obstructing masses to the small loose colonies and the single micrococci (Plate v., fig. 10).

In the compressed glomeruli colonies were very exceptionally present, and then only of small size. Isolated obstructing micrococcus masses were also present in the vascular capillary network of the medullary substance. They were also present in an isolated form in almost all the vessels. Accumulations of whitish corpuscles in the neighbourhood of the micrococci and alterations in the epithelial cells of the uriniferous tubules were not observed. The micrococci were never seen in the interior of the uriniferous tubules. The spleen contained loosely-arranged micrococcus colonies scattered about in the capillaries in moderate numbers, and also isolated dense deposits which distended the small vessels at the border and in the interior of the Malpighian corpuscles for short distances.

In the capillaries surrounding the intestinal glands numerous obstructing micrococcus masses were present (Plate i., fig. 11). At many points these were so extensive that branching accumulations were seen consisting entirely of these organisms.

The liver, like the lungs, contained no great accumulation of micrococci.

The largest diameter of an isolated micrococcus was ·8 to 1·0 mikrm.

These organisms differ from the micrococci of pyæmia very markedly as regards size, and in most other points. Thus they never enclose the blood corpuscles, even when they have accumulated in large numbers in the interior of the blood-vessels. They rather push them on one side. They do not cause coagulation of the blood, and thus emboli do not occur.

Only in one point do they resemble the pyæmic micrococci, namely, they do not show an increasing virulence when inoculated into a series of animals in succession.

Thus a syringeful (ten drops) of blood taken from the heart of the second rabbit was injected subcutaneously into a third rabbit. This animal died in thirty-six hours, the naked eye and microscopic appearances being exactly the same as in rabbit No. 2.

From rabbit No. 3 two drops of blood were injected into a mouse, and one drop into a rabbit.

The mouse died in thirty-seven hours; the rabbit remained unaffected.

The blood and all the organs of this mouse contained these oval micrococci, just as in the rabbit. A second mouse was then inoculated with blood taken from this mouse's heart, the operation being performed in the following manner:— The point of a scalpel was dipped in the blood of the heart, and about one-tenth of a drop was put into a small pocket-like wound on the root of the tail. This animal remained healthy.

On a second occasion I have, by injection of putrid meat infusion, obtained in a rabbit the same septicæmic process with the same oval micrococci.

Here, also, the disease could only be transmitted to other rabbits when at least five to ten drops of blood were used for inoculation.

6. Erysipelas in Rabbits.

Not only were large quantities of putrid fluids *injected* into rabbits, but various attempts were also made to produce disease by *inoculation* with different putrefying materials. These were not successful. In one case, however, after inoculation of the ear of a rabbit with mouse's dung softened in distilled water, redness and swelling occurred and spread slowly downwards from the point of inoculation. This redness extended on the fifth day as far as the root of the ear. When held up to the sunlight the ear which had not been inoculated appeared unaltered, only the chief blood-vessels being seen, while the inoculated ear, similarly illuminated, presented a uniformly dark red appearance, the individual vessels being no longer recognisable. It was thicker, and at the same time had become more fluid than the other, its point being bent and hanging down in consequence of its weight. The animal was, moreover, evidently ill, and died on the seventh day.

No effect was produced on another rabbit by injecting into it blood taken from the first. Unfortunately an attempt was not made to transmit the morbid process by inoculation with material from the ear of the affected rabbit to that of another.

Neither in the blood nor in the internal organs of this rabbit were any alterations found worthy of mention, notably no bacteria. The state of matters in the ear was, however, so remarkable, and bore so unmistakably the characters of a parasitic disease, that I have considered it right, although the infective nature of the disease was not directly demonstrated, to give here a description of the affection.

In transverse sections of the ear the blood-vessels were seen to be markedly dilated, full of red corpuscles, and surrounded by the nuclei of numerous white corpuscles. These nuclei were more numerous towards the cartilage of the ear, and on its surface they formed a pretty uniform dense layer. Between this layer and the cartilage cells were seen small fine rods arranged at pretty equal distances, which rods were distributed parallel to the cartilage in the dense cellular tissue which lies immediately outside the cartilage cells. In many places only single rods were seen; in others several were present, being arranged parallel to each other; while again, thickly interwoven clumps of these same rods were found, and that in parts where the white corpuscles were somewhat more thickly accumulated than elsewhere. These rods were present nowhere except close to the cartilage. Longitudinal sections were accordingly prepared which showed very distinctly the distribution of the rods on the surface of the cartilage. Figure 12, Plate i., is drawn from such a section. The large round bodies (r) are the nuclei of large flat cells, under which lie the cartilage cells. On this layer of flat cells there is a thick network consisting of bacilli, and outside these bacilli, indeed in fact covering them, are the nuclei (*b*), of the white corpuscles of which, however, only a small portion remains in the section. In many places the bacilli form more or less round dense clumps (as shown in fig. 12 *a*) which look like a pad of hair. From these clumps long rows of bacilli, in which the organisms become fewer and fewer, radiate in all directions. This reminds one of the peculiar, often starlike, figures which the bacillus anthracis forms when inoculated on the cornea of a living rabbit.* This network of bacilli extended over the whole cartilage of the ear on both surfaces. As the morbid process could be traced in its extension from its origin at the point of inoculation over the whole ear, and as throughout the whole limit of the process these bacilli were present, and as, further, the signs of inflammatory reaction were most marked in the immediate neighbourhood of these organisms, I consider it indubitable that they were the cause of the disease. I have never observed any formation of spores in them. They vary much in length. One rod, in which I could with certainty only distinguish two joints, was 3 mikrm. in length. The longest rod, consisting of six or seven joints, was 9—10 mikrm. in length. They are about ·3 mikrm. thick (the bacillus anthracis may be as long as 20 mikrm. and as thick as 1 to 1·25 mikrm.,—that is to say, about twice as long and three or four times as thick as the bacilli of the rabbit's ear).

* Frisch, *l. c.*, Pl. i., fig. 8, and Pl. ii., figs. 9 and 10.

ANTHRAX.

THE numerous investigations which have been made with regard to anthrax have almost all had reference to the behaviour of the bacillus anthracis outside the animal body. With regard to their numbers in the blood the conclusions have, as a rule, been drawn from blood taken by chance from any part of the body. No observations have yet been published as to the number of bacilli really present in the body, and as to their distribution in the vascular system.

In order to supply this deficiency, and because the bacillus anthracis behaves so like the septicæmic bacilli (thus being useful for comparison with the other pathogenic bacteria here described), I have examined rabbits and mice which have died of inoculated anthrax in the same manner as I did those killed by artificial traumatic infective diseases.

The staining of the bacillus alone, as obtained by treatment of sections stained in methyl violet with carbonate of potash, here proved of the greatest service. The mucous membrane of the stomach and intestine can, for example, be so prepared for examination by this method, that even with low powers the bacilli may be seen in all the vessels. In like manner sections of lung, liver, and kidneys, furnish extremely distinct and instructive preparations.

Although I had often previously examined the blood of animals suffering from anthrax, and had thus formed a high estimate as to the number of bacilli present in the body of an anthracic animal, yet I was quite surprised when I saw for the first time sections and portions of organs stained in this way, as *e.g.*, the intestinal mucous membrane and the iris of a rabbit. When magnified fifty diameters such a preparation presents at the first glance an appearance as if a blue colouring-matter had been injected into the vessels. Each intestinal villus is permeated by an exceedingly delicate blue network; in the mucous membrane of the stomach all the capillary network surrounding the gastric glands is stained blue; in the ciliary processes each projection is injected, and a spiral vessel stained of a dark blue colour leads from thence to the iris and breaks up into a fine blue network with loops directed towards the edge of the iris. The liver and lungs, and the glandular structures, such as the pancreas and salivary glands, are completeiy permeated by the same blue vascular network. Indeed there is no organ which is not more or less injected with the blue mass. It is, however, very striking that this injection is only present in the capillary vessels. All the larger vessels, even the arteries and veins of an intestinal villus, are either not at all stained or have but a light blue streak in their interior, and that only here and there. When magnified 250 times one can see that the blue capillary network is composed of numerous delicate rods (Plate iii., fig. 13), and when a power of 700 diameters is used it is found that the apparent injection is nothing more or less than the bacillus anthracis, stained dark blue, and present in incredible numbers in the whole capillary system. In the other vessels, especially in the larger ones, often only a single bacillus may be met with at long intervals, or they may be quite absent. Here, therefore, we have a striking proof of how little value are conclusions drawn in traumatic infective diseases from the examination of a drop of blood taken from a blood-vessel by chance; for one might quite well take a drop of blood from the heart and find no micro-organisms in it, or one might readily overlook the few which might be present, and that although the capillary system abounds in them.

The distribution of the bacillus anthracis in the capillaries is not, however, quite uniform. There are fewer in the brain, in the skin, in the capillaries of the muscle, and in the tongue than elsewhere; on the other hand, in the liver, lungs, kidneys, spleen, intestines, and stomach, they are always present in the enormous numbers above described. The spleen, to which the disease owes its name, is not distinguished above the other organs which have been mentioned by a greater number of bacilli. In the capillaries themselves the bacilli accumulate in largest numbers at the point farthest removed from the nearest afferent artery and the efferent vein;

that is to say, where the arterial capillaries join the venous, the aggregate diameter of the vessels being there the broadest and the blood stream flowing the slowest. In the intestinal villus this spot is at its apex and the neighbouring part of the periphery; in the liver it lies midway between the ultimate twigs of the hepatic vein and of the vena portæ. Among the points where the bacilli accumulate in greatest numbers are also the glomeruli of the kidney, which become for the most part transformed into clumps of bacilli. It by no means unfrequently happens that from the presence of the rapidly increasing bacilli at the places mentioned, chiefly in the glomeruli, intestinal villi, mucous membrane of the stomach, salivary glands, and pancreas, individual capillaries become torn, and blood with the contained bacilli is extravasated. This occurs most frequently in the glomeruli. Many of these burst and the bacilli pass into the uriniferous tubules. They do not, however, extend far into these; at least I have only found them in the commencement of the convoluted tubules, in which they form long threads interwoven with each other. I have never seen bacilli in the straight uriniferous tubules.

The facts which I have described are those met with in rabbits. Mice, which I have often investigated, behave essentially in the same manner. In the latter animals, however, the spleen is more especially the seat of the bacilli; then come the lungs, and last of all the kidneys. The contrast between the very large numbers of bacilli in the capillaries and their small quantity in the large vessels is even more striking in the mouse than in the rabbit.

I have also had an opportunity of examining the lungs, liver, spleen, and kidneys of sheep which had died of anthrax, and I found here also that the bacilli had the same relations as regards numbers and distribution as in the rabbit.

I would recommend the study of organs taken from animals affected with anthrax, and stained in the manner described, to those who, in spite of all the proof already furnished, do not yet regard it as a parasitic disease. The simple fact that death occurs in twenty-four hours after inoculation with the smallest drop of anthracic blood, provided that it contains bacilli or their spores, and that then almost all the capillaries of the lungs, liver, kidneys, spleen, intestine, stomach, &c. (placed in absolute alcohol immediately after death), are found to be filled with enormous numbers of the same bacilli, has so evidently only one interpretation that no commentary is required. The investigator who still considers the presence of these organisms as accidental, quite immaterial, or merely accessory, must, before he can attribute the death of the animal to some other unknown ferment, consider likewise as immaterial the loss of the constituents of the blood which go to build up these innumerable bacilli, the accumulation of waste products which such a rapid interchange of material as the growth of the bacilli must of necessity involve, and also the disturbances in the circulation and in the nourishment of important organs induced by the plugging of most of the capillaries. But in that case there would be no reason why, in the case of trichinosis, scabies, and other parasitic diseases communicated by direct contact, some specific ferments in addition to trichinæ, acari, &c., should not also be supposed to be present.

CONCLUSIONS.

I AM well aware that the investigations above described are very imperfect. It was necessary, in order to have time for those parts of the investigation which seemed the most important and essential, to omit the examination of many organs, such as the brain, heart, retina, &c., which ought not to pass unnoticed in researches on infective diseases. For the same reason no record was kept of the temperature, although this would undoubtedly have yielded most interesting results. I have intentionally refrained from entering into details of morbid anatomy, as only the etiology interested me, and as I did not feel qualified to undertake a study of the morbid anatomy of traumatic infective diseases. I must therefore leave this part of the investigation to those who are better able to undertake it.

Nevertheless I consider that the results of my researches are sufficiently definite to enable me to deduce from them some well-founded conclusions.

In this summary I shall, however, confine myself to the most obvious conclusions. It has indeed of late become too common to draw the most sweeping conclusions as to infective diseases in general from the most unimportant observations on bacteria. I shall not follow this custom, although the material at my command would furnish rich food for meditation. For the longer I study infective diseases the more am I convinced that generalisations of new facts are here a mistake, and that every individual infective disease or group of closely allied diseases must be investigated for itself.

As regards the artificial traumatic infective diseases observed by me, the conditions, which must be established before their parasitic nature can be proved, were completely fulfilled in the case of the first five, but only partially in that of the sixth. For the infection was produced by such small quantities of fluid (blood, serum, pus, &c.) that the result cannot be attributed to a merely chemical poison.

In the materials used for inoculation bacteria were without exception present, and in each disease a different and well-marked form of organism could be demonstrated.

At the same time, the bodies of those animals which died of the artificial traumatic infective diseases contained bacteria in such numbers that the symptoms and the death of the animals were sufficiently explained. Further, the bacteria found were identical with those which were present in the fluid used for inoculation, and a definite form of organisms corresponded in every instance to a distinct disease.

These artificial traumatic infective diseases bear the greatest resemblance to human traumatic infective diseases, both as regards their origin from putrid substances, their course, and the result of post-mortem examination. Further, in the first case, just as in the last, the parasitic organisms could be only imperfectly demonstrated by the earlier methods of investigation; not till an improved method of procedure was introduced was it possible to obtain complete proof that they were parasitic diseases. We are therefore justified in assuming that human traumatic infective diseases will in all probability be proved to be parasitic when investigated by these improved methods.

On the other hand, it follows from the fact that a definite pathogenic bacterium, *e.g.*, the septicæmic bacillus, cannot be inoculated on every variety of animal (a similar fact is also true with regard to the bacillus anthracis); that the septicæmia of mice, rabbits, and man are not under all circumstances produced by the same bacterial form. It is of course possible that one or other of the bacteric forms found in animals also play a part in such diseases in the human subject. That, however, must be specially demonstrated for each case; *a prïori* one need only expect that bacteria are present; as regards form, size, and conditions of growth, they may be similar, but not always the same, even in what appear to be similar diseases in different animals.

Besides the pathogenic bacteria already found in animals, there are no doubt many others. My experiments refer only to those diseases which ended fatally. Even these are in all probability not exhausted in the six forms mentioned. Further experiments on many different species of animals, with the most putrid substances and with every possible modification in the method of application, will doubtless bring to light a number of other infective diseases, which will lead to further conclusions regarding infective diseases and pathogenic bacteria.

But even in the small series of experiments which I was able to carry out, one fact was so prominent that I must regard it as constant, and, as it helps to remove most of the obstacles to the admission of the existence of a contagium vivum for traumatic infective diseases, I look on it as the most important result of my work. I refer to the differences which exist between pathogenic bacteria and to the constancy of their characters. A distinct bacteric form corresponds, as we have seen, to each disease, and this form always remains the same, however often the disease is transmitted from one animal to another. Further, when we succeed in reproducing the same disease *de novo* by the injection of putrid substances, only the same bacteric form occurs which was before found to be specific for that disease.

Further, the differences between these bacteria are as great as could be expected between particles which border on the invisible. With regard to these differences, I refer not only to the size and form of the bacteria, but also to the conditions of their growth, which can be best recognised by observing their situation and grouping. I therefore study not only the individual alone, but the whole group of bacteria, and would, for example, consider a micrococcus which in one species of animal occurred only in masses (*i.e.*, in a zooglœa form), as different from another which in the same variety of animal, under the same conditions of life, was only met with as isolated individuals. Attention must also be paid to the physiological effect, of which I scarcely know a more striking example than the case of the bacillus and the chain-like micrococcus growing together in the cellular tissue of the ear; the one passing into the blood and penetrating into the white blood corpuscles, the other spreading out slowly in the tissue in its vicinity and destroying everything round about; or again, the case of the septicæmic and pyæmic micrococci of the rabbit in their different relations to the blood; or lastly, the bacilli extending only over the surface of the aural cartilage in the erysipetalous disease, as contrasted with the bacillus anthracis likewise inoculated on the rabbit's ear, but quickly passing into the blood.

As, however, there corresponds to each of the diseases investigated a form of bacterium distinctly characterised by its physiological action, by its conditions of growth, size, and form, which, however often the disease be transmitted from one animal to another, always remains the same and never passes over into other forms, *e.g.*, from the spherical to the rod-shaped, we must in the meantime regard these different forms of pathogenic bacteria as distinct and constant species.

This is, however, an assertion which will be much disputed by botanists, to whose special province this subject really belongs.

Amongst those botanists who have written against the subdivision of bacteria into species, is Nägeli, who says,* "I have for ten years examined thousands of different forms of bacteria, and I have not yet seen any absolute necessity for dividing them even into two distinct species."

Brefeld † also states that he can only admit the existence of specific forms justifying the formation of distinct species when the whole history of development has been traced by cultivation from spore to spore in the most diverse nutritive fluids.

Although Brefeld's demand is undoubtedly theoretically correct, it cannot be made a *sine quâ non* in every investigation on pathogenic bacteria. We should otherwise be compelled to cease our investigations into the etiology of infective diseases till botanists have succeeded in finding out the different species of bacteria by cultivation and development from spore to spore. It might then very easily happen that the endless

* Die Niederen Pilze. München, 1877, p, 20.

† Untersuchungen über die Spaltpilze. Sitzungsbericht der Gesellschaft Naturforschender Freunde in Berlin. 19th Feb., 1878.

trouble of pure cultivation would be expended on some form of bacterium which would finally turn out to be scarcely worthy of attention. In practice only the opposite method can work. In the first place certain peculiarities of a particular form of bacterium different from those of other forms, and in the second place its constancy, compel us to separate it from others less known and less interesting, and provisionally to regard it as a species. And now, to verify this provisional supposition, the cultivation from spore to spore may be undertaken. If this succeeds under conditions which shut out all sources of fallacy, and if it furnishes a result corresponding to that obtained by the previous observations, then the conclusions which were drawn from these observations and which led to its being ranked as a distinct species must be regarded as valid.

On this, which as it seems to me is the only correct practical method, I take my stand, and, till the cultivation of bacteria from spore to spore shows that I am wrong, I shall look on pathogenic bacteria as consisting of different species.

In order, however, to show that I do not stand alone in this view, I shall here mention the opinion of some botanists who have already come to a similar conclusion.

Cohn* states that, in spite of the fact that many dispute the necessity of separating bacteria into genera and species, he must nevertheless adhere to the method as yet followed by him, and separate bacteria of different form and fermenting power from each other, so long as complete proof of their identity is not given.

From his investigations on the effects of different temperatures and of dessication on the development of bacterium termo, Eidam † came to the conclusion that different forms of bacteria require different conditions of nutriment, and that they behave differently towards physical and chemical influences. He regards these facts as a further proof of the necessity of dividing organisms into distinct species.

I shall bring forward another reason to show the necessity

* Beiträge zur Biologie der Pflanzen. Bd. I., Heft. 3, p. 144.

† Beiträge zür Biologie der Pflanzen. Bd. I., Heft. 3, p. 223.

of looking on the pathogenic bacteria which I have described as distinct species. The greatest stress, in investigations on bacteria, is justly laid on the so-called pure cultivations, in which only one definite form of bacterium is present. This evidently arises from the view that if, in a series of cultivations, the same form of bacterium is always obtained, a special significance must attach to this form: it must indeed be accepted as a constant form, or, in a word, as a species. Can, then, a series of pure cultivations be carried out without admixture of other bacteria? It can in truth be done, but only under very limited conditions. Only such bacteria can be cultivated pure, with the aids at present at command, which can always be known to be pure, either by their size and easily recognisable form, as the bacillus anthracis, or by the production of a characteristic colouring matter, as the pigment bacteria. When, during a series of cultivations, a strange species of bacteria has by chance got in, as may occasionally happen under any circumstances, it will in these cases be at once observed, and the unsuccessful experiment will be thrown out of the series without the progress of the investigation being thereby necessarily interfered with.

But the case is quite different when attempts are made to carry out cultivations of very small bacteria, which, perhaps, cannot be distinguished at all without staining; how are we then to discover the occurrence of contamination? It is impossible to do so, and therefore all attempts at pure cultivation in apparatus, however skilfully planned and executed, must, as soon as small bacteria with but little characteristic appearances are dealt with, be considered as subject to unavoidable sources of fallacy, and in themselves inconclusive.

But nevertheless a pure cultivation is possible, even in the case of the bacteria which are smallest and most difficult to recognise. This, however, is not conducted in cultivation apparatus, but in the animal body. My experiments demonstrate this. In all the cases of a distinct disease, *e. g.*, of septicæmia of mice, only the small bacilli were present, and no other form of bacterium was ever found with it, unless in the case where that causing the tissue gangrene was intentionally inoculated at the same time. In fact, there exists no better cultivation apparatus for pathogenic bacteria than the animal body itself. Only a very limited number of bacteria can grow in the body, and the penetration of organisms into it is so difficult that the uninjured living body may be regarded as completely isolated with respect to other forms of bacteria than those intentionally introduced. It is quite evident, from a careful consideration of the two diseases produced in mice—septicæmia and gangrene of the tissue—that I have succeeded in my experiments in obtaining a pure cultivation. In the putrefying blood, which was the cause of these two diseases, the most different forms of bacteria were present, and yet only two of these found in the living mouse the conditions necessary for their existence. All the others died, and these two alone, a small bacillus and a chain-like micrococcus, remained and grew. These could be transferred from one animal to another as often as was desired, without suffering any alteration in their characteristic form, in their specific physiological action and without any other variety of bacteria at any time appearing. And further, as I have demonstrated, it is quite in the power of the experimenter to separate these two forms of bacteria from each other. When the blood in which only the bacilli are present is used, these alone are transmitted, and thenceforth are obtained quite pure; while on the other hand, when a field-mouse is inoculated with both forms of bacteria, the bacilli disappear, and the micrococcus can be then cultivated pure. Doubtless an attempt to unite these two forms again in the same animal by inoculation would have been successful. In short, one has it completely in one's power to cultivate several varieties of bacteria together, to separate them from each other, and eventually to combine them again. Greater demands can hardly be made on a pure cultivation, and I must therefore regard the successive transmission of artificial infective diseases as the best and surest method of pure cultivation. And it can further claim the same power of demonstrating the existence of specific forms of bacteria, as must be conceded to any faultless cultivation experiments.

From the fact that the animal body is such an excellent apparatus for pure cultivation, and that, as we have seen, when the experiments are properly arranged and sufficient optical aids used, only one specific form of bacterium can be found in each distinct case of artificial traumatic infective disease, we may now further conclude that when, in examining a traumatic infective disease, several different varieties of bacteria are found, as *e.g.*, chains of small granules, rods, and long oscillating threads (such as were seen together by Coze and Feltz in the artificial septicæmia of rabbits, see p. 9), we have to do with either a combined infective disease,—that is, not a pure one,—or, what in the case cited is more probable, an inexact and inaccurate observation. When, therefore, several species of bacteria occur together in any morbid process, before definite conclusions are drawn as to the relations of the disease in question to the organisms, either proof must be furnished that they are all concerned in the morbid process, or an attempt must be made to isolate them and to obtain a true pure cultivation. Otherwise we cannot avoid the objection that the cultivation was not pure, and therefore not conclusive. I shall only briefly refer to a further necessary consequence of the admission of the existence of different species of pathogenic bacteria. The number of the species of these bacteria is limited; for, of the numerous diverse forms present in putrid fluids, one or but few can in the most favourable cases develop in the animal body. Those which disappear are, for that species of animal at least, not pathogenic bacteria. If, however, as follows from the foregoing, there exist hurtful and harmless bacteria, experiments performed on animals with the latter, *e. g.*, with bacterium termo, prove absolutely nothing for or against the behaviour of the former—the pathogenic—forms. But almost all the experiments of this nature have been carried out with the first mixture of different species of bacteria which came to hand without there being any certainty that pathogenic bacteria were in reality present in the mixture. It is therefore evident that none of these experiments can be regarded as furnishing evidence of any value for or against the parasitic nature of infective diseases.

In all my experiments, not only have the form and size of the bacteria been constant, but the greatest uniformity in their actions on the animal organism has been observed, though no increase in virulence, as described by Coze and Feltz, Davaine, and others. This leads me to make some remarks on the supposed law of the increasing virulence of blood when transmitted through successive animals, discovered or confirmed by the investigators just mentioned.

The discovery of this law has, as is well known, been received with great enthusiasm, and it has excited no little interest owing to its intimate bearing on the doctrine of natural selection (Anpassung und Vererbung). Some investigators, who are in other things very exact, have allowed themselves to be blinded by the seductive theory that the insignificant action of a single putrefactive bacterium may, by continued natural selection in passing from animal to animal, be increased in virulence till it becomes deadly though a drop of the infective liquid be diluted a quadrillion times. They have founded thereon the most beautiful practical applications, not suspecting that the bacteria in question have never been with certainty demonstrated.

The original works of Coze and Feltz, as also that of Davaine, are not at my disposal for reference; and I cannot therefore enter into a complete criticism of them. So far, however, as I can gather from the references accessible to me, especially from the detailed notices in Virchow and Hirsch's 'Jahresbericht,' no complete proof that the virulence of septicæmic blood increases from generation to generation seems to have been furnished. Apparently blood more and more diluted was injected, and astonishment was felt when this always acted, the effect being then ascribed to its increasing virulence. But controlling experiments to ascertain whether the septicæmic blood were not already as virulent in the second and third generations as in the twenty-fifth, do not seem to have been made. My experiments so far support and are in accordance with those of Coze, Feltz, and Davaine in that for the first infection of an animal relatively large quantities of putrid fluids are necessary; but in the second generation, or at the latest in the third, the full virulence was attained, and afterwards remained constant.

Of my artificial infective diseases the septicæmia of the mouse has the greatest correspondence with the artificial septicæmia described by Davaine. If we were to experiment with this disease in the same manner as Davaine experimented, we would, if no controlling experiments were employed, find the same increase in virulence of the disease. It would only be necessary to use blood in slowly decreasing quantities in order to obtain in this way any progressive increase of the virulence that might be desired. I, however, took from the second or third animal the smallest possible quantity of material for inoculation, and thus arrived more quickly at the greatest degree of virulence. Till, therefore, I am assured that, in the septicæmia observed by Davaine, such controlling experiments were made, I can only look on an increase in virulence as holding good for the earlier generations. In order to explain this we do not, however, require to have recourse to the magical wand of natural selection; a feasible explanation can be very naturally furnished. Let us take again the septicæmia of mice, as being the most suitable example.

If two drops of putrefying blood be injected into such an animal there is introduced not only a number of totally distinct species of bacteria, but also a certain amount of dissolved putrid poison (sepsin), not sufficient to produce a fatal effect, but yet certainly not without influence on the health of the animal. Different factors must therefore be considered as affecting the health of the animal. On the one hand there is the dissolved poison, on the other the different species of bacteria, of which, however, perhaps only two, as in the example before us, can multiply in the body of the mouse and there exert a continuous noxious influence. Only one of these two species can penetrate into the blood, and if the blood alone be used for further inoculations, only this one variety will come victorious out of the battle for existence. The further development of the experiment depends entirely on the quantity of the putrid poison, and on the relation of the two forms of bacteria to each other in point of numbers. If one injects a large amount of the septic poison and a large number of that variety of bacteria which increases locally (in this case the chain-like micrococci causing the gangrene of the tissue), but only a very small number of the bacteria which pass into the blood (here the bacilli), the first animal experimented on will die, as a result of the preponderating influence of the first two factors before many bacilli can have got into the blood and multiplied there. Of the blood of this first animal, containing, as it does proportionately very few bacilli, one-fifth to one-tenth of a drop must be inoculated in order to convey the disease with certainty. In the second animal, however, only the bacilli are introduced, and these develop undisturbed in the blood. For the infection of the third animal the smallest quantity of this blood which can produce an effect is then sufficient, and after this third generation the virulence of the blood remains uniform.

We may also imagine another case in which the increase of the virulence may go on through more than two generations without any modification resulting from natural selection and transmission from animal to animal. This would take place if several species of bacteria capable of passing into the blood were introduced into the animal at the first injection. Let us suppose, for example, that in the same putrefying blood which served for the foregoing experiment, the bacilli of anthrax were also present, there would be then contained in the blood of the first animal not only the septicæmic bacillus, but also bacillus anthracis, and of each only a small number; of the anthrax bacillus there would be even fewer than of the other, because in mice they are deposited chiefly in the spleen, lungs, &c.; while in the blood of the heart they are, even in the most favourable cases, only sparsely distributed. On the other hand, the anthrax bacilli have this advantage, that, provided they be inoculated in considerable numbers, they kill even within twenty hours, while the septicæmic bacilli only destroy life after fifty hours. In the blood of the second animal, therefore, both species of bacilli would be present in larger numbers than in the first, although not yet so numerous as if either organism had been inoculated singly. Hence a larger quantity of blood is necessary to ensure transmission to a third animal. Perhaps this might be the case even in the fourth generation, till finally one or other variety of bacillus would alone be present in the blood injected. Probably this would be the septicæmic bacillus.

In this way the experiments of Coze, Feltz, and Davaine may admit of simple explanation and be brought into harmony with my results.

PRINTED BY WEST, NEWMAN AND CO., 54, HATTON GARDEN, LONDON, E.C.

EXPLANATION OF THE PLATES.

All the drawings have been made as true to nature as possible by the use of the camera lucida and of Zeiss's $\frac{1}{12}$ inch oil immersion objective. An object micrometer was used to determine the magnifying power.

PLATE I.

Fig. 5.—Longitudinal section of the ear of a mouse. Progressive gangrene of the tissue:—

a. Normal cartilage and on each side normal tissue.
b. Line of demarcation, accumulation of nuclei.
c. Gangrenous portion of the ear devoid of nuclei.
d. Transverse section of a vessel full of micrococci. × 25.

Fig. 8. — Marginal zone of a cheesy abscess in a rabbit. Lateral portion:—

a. Cloud-like masses of zooglœa.
b. and *c.* Smaller.
d. Smallest micrococcus colonies.
e. Accumulation of nuclei in the neighbourhood of the zooglœa.
f. Broken-down nuclei.
g. Dead portion of zooglœa. × 700.

Fig. 11.—Capillary vessel from the mucous membrane of the small intestine of a septicæmic rabbit:—

a. Nuclei of the wall of the vessel.
b. Oval micrococci. × 700.

PLATE II.

Fig. 1.—Blood of a septicæmic mouse, dried on a cover glass, stained with methyl violet, and mounted in Canada balsam. Red blood corpuscles are seen, and among them are small bacilli. × 700.

Fig. 2.—White blood corpuscles from one of the veins of the diaphragm of a septicæmic mouse. All stages of transition are shown from blood corpuscles which contain but few bacilli to those which have become converted into masses of bacilli. × 700.

Fig. 4.—Blood of a mouse affected with anthrax. Red blood corpuscles and anthrax bacilli. The specimen was prepared in the same way as that shown in fig. 1. The points of junction of the bacilli are too strongly drawn. × 700.

Fig. 12.—Section of the ear of a rabbit parallel to the surface of the cartilage. The morbid process resembled erysipelas :—

a. Ball-like accumulation of bacilli.

b. Accumulation of nuclei above the layer of bacilli.

c. Nuclei of flat cells connected with the cartilage below the layer of bacilli.

d. Bacilli arranged parallel to each other. × 700.

PLATE III.

FIG. 8.—Vein of the diaphragm of a septicæmic mouse:—

a. Nuclei of the vascular wall.

b. Septicæmic bacilli.

c. White blood corpuscles which have become transformed into masses of bacilli.

d. Capillaries opening into the vein. × 700.

FIG. 7.—Marginal zone of a cheesy abscess cavity in a rabbit. Lower surface:—

a. Accumulation of nuclei at the outer border of the abscess.

b. Zooglœa, consisting of very small micrococci (these have been drawn too large in parts, more especially in the interior of the zooglœa).

c. Zooglœa, partially dead.

d. Dead zooglœa.

e. Remnants of nuclei. × 700.

Fig. 3

PLATE IV.

Fig. 6.—A portion of the cartilage and adjacent tissue in the vicinity of *c*, fig. 5. Magnified 700 diameters:—

a. Necrotic cartilage cells.

b. Chain-like micrococci in masses.

c. The same isolated. × 700.

Fig. 13.—Villus of a rabbit. Anthrax. The bacilli alone stained. × 250.

PLATE V.

Fig. 9.—Vessel from the cortex of the kidney of a pyæmic rabbit:—

a. Nuclei of the vascular wall.
b. Small group of micrococci between blood corpuscles.
c. Dense masses of micrococci adherent to the wall and enclosing blood corpuscles.
d. Pairs of micrococci at the border of the large mass. × 700.

Fig. 10.—Glomerulus of a septicæmic rabbit:—

a. Capillary loop with oval micrococci spread out like a membrane.
b. Micrococci deposited on the walls of a capillary vessel.
c. Loop completely filled with micrococci.
d. Individual micrococci in a capillary vessel near a glomerulus. × 700.

Fig. 14.—A part of the vascular network of the same. × 700.

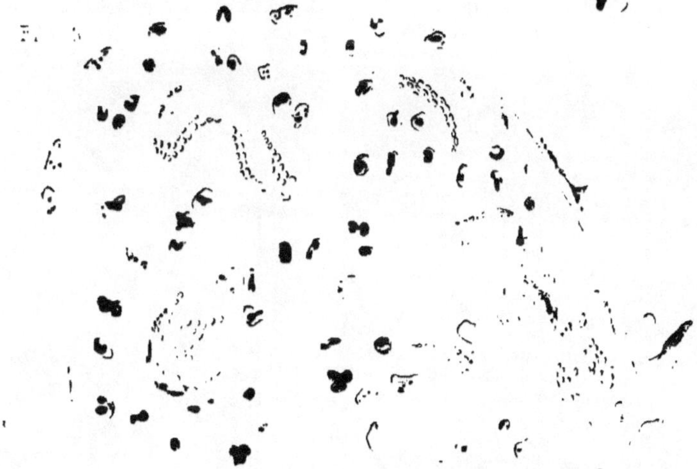

REPORT

PRESENTED TO THE

TWENTY-SECOND ANNUAL MEETING

OF THE

NEW SYDENHAM SOCIETY

HELD AT CAMBRIDGE,

AUGUST 8TH, 1880.

WITH

Classified List of Published Works

AND OTHER INFORMATION.

OFFICERS FOR 1880-81.

President.

SIR WILLIAM W. GULL, Bart., M.D., F.R.S., D.C.L., LL.D.

Vice-Presidents.

ROBERT BARNES, M.D.
SIR GEORGE BURROWS, F.R.S., Bart.
R. W. FALCONER, M.D., D.C.L. (Bath).
W. D. HUSBAND, Esq. (York).
W. T. GAIRDNER, M.D. (Glasgow).
CÆSAR H. HAWKINS, Esq., F.R.S.
T. HOLMES, Esq.
J. HUGHLINGS JACKSON, M.D., F.R.S.
*GEORGE JOHNSON, M.D., F.R.S.
JOSEPH LISTER, Esq., F.R.S.
SIR JAMES PAGET, F.R.S., Bart.
*GEORGE PAGET, M.D., F.R.S. (Cambridge).
*WILLIAM RUTHERFORD, M.D., F.R.S. (Edinburgh).
T. GRAINGER STEWART, M.D. (Edinburgh).
SIR THOMAS WATSON, M.D., F.R.S., Bart.
HERMANN WEBER, M.D.

Council.

*T. CLIFFORD ALLBUTT, M.D., F.R.S.
*MILNER BARRY, M.D. (Tunbridge Wells).
THOMAS BARLOW, M.D.
*LAUDER BRUNTON, M.D., F.R.S.
W. H. BROADBENT, M.D.
THOMAS BUZZARD, M.D.
W. CHOLMELEY, M.D.
WILLIAM COLLES, M.D. (Dublin).
R. M. CRAVEN, Esq. (Hull).
J. LANGDON H. DOWN, M.D.
DYCE DUCKWORTH, M.D.
J. MATTHEWS DUNCAN, M.D.
JOHN EASTON, M.D.
SIR JOSEPH FAYRER, M.D.
C. J. HARE, M.D.
G. E. HERMAN, M.D.
H. MACNAUGHTEN JONES, M.D. (Cork).
J. C. LANGMORE, M.B.
P. W. LATHAM, M.D. (Cambridge).
*STEPHEN MACKENZIE, M.D.
J. W. MOORE, M.D. (Dublin).
*WALTER MOXON, M.D.
*WILLIAM ROBERTS, M.D. (Manchester).
SEPTIMUS W. SIBLEY, Esq.
J. K. SPENDER, M.D. (Bath).
*WILLIAM SQUARE, Esq. (Plymouth).
*PAUL SWAYNE, Esq. (Devonport).
T. P. TEALE, Esq. (Leeds).
WILLIAM TURNER, M.B., F.R.S.E. (Edinburgh).
JOHN WALTERS, M.B. (Reigate).
JAMES WEST, Esq. (Birmingham).

Treasurer.

W. SEDGWICK SAUNDERS, M.D., 13, Queen Street, Cheapside, E.C.

Auditors.

E. CLAPTON, M.D.
S. FENWICK, M.D.
F. M. CORNER, Esq.

Hon. Secretary.

JONATHAN HUTCHINSON, Esq., 15, Cavendish Square, W.

Those marked with an Asterisk were not in office last year.

REPORT

Presented to the Twenty-second Annual Meeting of the New Sydenham Society.

The series for the year 1879 consisted of the following works:—The second and concluding volume of Waring's Bibliotheca Therapeutica, a second part of the Society's Lexicon of Medical Terms, Guttmann's Manual of Physical Diagnosis, and a second Fasciculus of the Society's Atlas of Pathology. With the latter were included essays on the present state of knowledge as to the Pathology of the Kidney, by Dr. Greenfield, and as to that of the Spleen and Supra-Renals, by Dr. Goodhart. These papers were compiled at the request of the Council, and were freely illustrated by drawings from the microscope.

The series for the current year will probably comprise the following:—

I. A third fasciculus of the Society's Lexicon. (Already out).

II. The fifth and concluding volume of Hebra's Treatise on Skin Diseases, with index to the whole. (Already out).

III. Koch's Researches on the Etiology of diseases consequent on Wound Infection.

IV. A third Fasciculus of the Society's Atlas of Pathology, comprising diseases of the Liver.

V. A fourth Fasciculus of the Lexicon.

The Council has adopted, for reprinting, the classical Treatise of Dr. Stokes on Diseases of the Chest. This work, which has been always held in very high estimation by all authorities, has been for some time out of print. It will be edited for the Society, with short annotations, &c.,

by Dr. Hudson, of Dublin. It has also been decided to edit for the Society a selection from the works of Duchenne; and Dr. Vivian Poore has, at the Council's request, undertaken the preparation of the work.

The translation of Professor Charcot's Lectures on the Diseases of Old Age, and on certain Chronic Maladies, has been decided on. The work has been placed in the hands of Mr. William S. Tuke.

The preparation of the Society's Lexicon is, in the hands of its editors, Mr. Power and Dr. Sedgwick, progressing satisfactorily and as rapidly as the difficulties of the task permit. Three Fasciculi have been issued, and another is just ready. It is to be distinctly understood that the Fasciculi of this work are always issued as soon as ready. The Council is prepared to devote to it any portion of the year's income that may be requisite; and nothing but the onerous nature of the editors' task will be allowed to delay its publication.

The Balance Sheet has been audited, and is, as usual, appended.

BALANCE SHEET FOR 1879.

Receipts.

	£	s.	d.	£	s.	d.
Balance in hand, Dec., 1878 (see preceding Balance Sheet)				1082	13	1¼
Subscriptions, 1 for 1862	1	1	0			
,, 1 ,, 1863	1	1	0			
,, 3 ,, 1864	3	3	0			
,, 1 ,, 1865	1	1	0			
,, 3 ,, 1866	3	3	0			
,, 2 ,, 1867	2	2	0			
,, 1 ,, 1868	1	1	0			
,, 2 ,, 1869	2	2	0			
,, 2 ,, 1870	2	2	0			
,, 3 ,, 1871	3	3	0			
,, 2 ,, 1872	2	2	0			
,, 2 ,, 1873	2	2	0			
,, 5 ,, 1874	5	5	0			
,, 9 ,, 1875	9	9	0			
,, 79 ,, 1876	82	19	0			
,, 136 ,, 1877	142	16	0			
,, 726 ,, 1878	762	6	0			
,, 1678 ,, 1879	1761	18	0			
,, 23 ,, 1880	24	3	0			
Volumes of Trousseau	128	12	0			
Fascic. of Atlas	12	12	0			
Other volumes	79	0	4			
	3033	3	4			
Less deductions per Local Secretaries	27	15	2			
				3005	8	2
				£4088	1	3¼

Expenditure.

	£	s.	d.	£	s.	d.
Folio I. Artists, editors, and translators				692	3	6
,, II. Printers				838	2	8
,, III. Paper				408	0	0
,, IV. Bookbinders				339	13	10
,, V. Expenses of Management:—						
Agent's Salary and Percentage	264	14	7			
Secretary's Expenses	55	13	0			
Treasurer's Expenses	4	7	3			
Insurance	12	0	0			
Agent's Expenditure (chiefly carriage)	84	9	10			
Advertisements	21	14	10			
				442	19	6
Sum Total of Expenses				2720	19	6
Balance in hand, December 31st, 1879				1367	1	9¼
				£4088	1	3¼

CLASSIFIED LIST

OF THE

SOCIETY'S PUBLICATIONS.

Medicine.

ON THE TEMPERATURE IN DISEASE: A MANUAL OF MEDICAL THERMOMETRY. By Dr. C. A. WUNDERLICH. (Leipzig). Translated by Dr. BATHURST WOODMAN. With forty Woodcuts and seven Lithographs.

"It is well to recollect that this contains not only observations on the temperature in disease, but also in health, and is a complete epitome as to the history of the subject up to date. It is a work of reference absolutely necessary for all who would keep themselves abreast of the day in relation to so important a matter as corporeal temperature."—*Edin. Med. Journ.*, May, 1872.

"In short, without pledging ourselves to Wunderlich as infallible, we may say, emphatically, that his is a masterwork, in which every part of his subject is considered with that thoroughness which comes of ripe knowledge and reflection. Let us add that Dr. Bathurst Woodman, following one or two laudable examples that have been set by other translators for the Sydenham Society, has enriched the work with notes of his own observations and those of other English writers, which are of no small value, and unquestionably do much to make the volume complete and full." - *Lancet*, April 20, 1872.

"The translator has rendered into readable English, and enriched with practical notes, a book which, even in its original form, has started into active work many physicians in England, France, and America, and which now, in its popular form, must render the diagnosis of disease infinitely more accurate."—*Medical Times and Gazette*, June 3, 1871.

"The publication of this volume marks an epoch in the history of medical thermometry. The very possibility of such a book—full not only of exact knowledge, but of important generalisations—is an indication that the great problems relating to the alterations in the human temperature—the problems of fever and collapse—are now being studied in a manner calculated to throw light on the hidden processes of disease. . . . . The value of this great work of Professor Wunderlich is that it lays open his vast clinical experience of the thermometer, and that it sketches in general terms the course of the temperature in various forms of disease."—*Glasgow Medical Journal*, August, 1871.

"This treatise displays so much perseverance and thoroughness, such admirable caution and insight, and such wide and minute learning, that it may be said not only to establish this branch of investigation for the first time upon a deep and lasting basis, but also to build up a very great part of the edifice, and to point out with clearness the directions in which future labour must be applied."--Dr. Allbutt in *Brit. and For. Med. Chir. Rev.*, April, 1870.

LECTURES ON CLINICAL MEDICINE, delivered at the Hotel Dieu, Paris. By Professor TROUSSEAU. Five Volumes. Vol. 1, translated, with notes and appendices, by the late

Dr. Bazire. Vols. 2 to 5, translated from the third edition, revised and enlarged, by Sir John Rose Cormack.

"We are indebted to the New Sydenham Society for this rich contribution to our medical literature. . . . . . Trousseau is an author to be read rather than reviewed. He can only be criticised worthily at the bedside. . . . . . We commend this great physician's work to the study of every reader."—*Lancet*, October 15, 1870.

"The above-mentioned works constitute the nineteenth annual issue to its subscribers of the New Sydenham Society; and, though relating to different subjects, we have classed them together, because it seems of more importance to the profession that they should know the very valuable practical information they can secure for one guinea, than at this time of day they should be treated to an elaborate critique on Trousseau's Clinical Medicine, or on Wunderlich's Treatise on Thermometry; the worth of these volumes being well known to all but the merest tyro in medicine."—*Edinburgh Medical Journal*, May, 1872.

"We should think any medical library absurdly incomplete now which did not have, alongside of Watson and Graves and Tanner, the clinical medicine of Trousseau. The work is full of the results of the richest natural observation, and is the production of one who was enlightened enough to combine with new methods of investigation the vigorous and independent ideas of the old physicians, whom he so eloquently magnifies. The volume is an extremely rich and valuable addition to the library of physicians and practitioners generally."—*Lancet*, December 4, 1869.

LATHAM'S COLLECTED WORKS. 2 vols. Edited by Dr. Robert Martin. With Memoir of Latham by Sir Thomas Watson.

"It indicates discrimination and taste on the part of those who conduct the New Sydenham Society, that they have selected for publication a work so different in many respects from the ephemeral books which issue in such numbers from the teeming press of the present day. . . . . . This is one of the few books which deserve to live, because it is full of real and conscientious work,—of observations, carefully, reverently, and modestly made during a long series of years,—of thoughts pondered and repondered with candour and self-distrust and willingness to be taught, while the literary execution is unmistakably that of a man of education, culture, and taste."—*Edinb. Med. Jour.*, March, 1877.

"The different subjects are dealt with in a way which will always render them fresh to the reader from tbe peculiarly original bent of the writer's mind, and the acuteness of his reasoning. We quite agree with the editor that 'where all are so admirable, it were perhaps well to avoid the singling out of any one as though pre-eminently good.' If we made any exception to this, it would be to specially direct attention to the articles on 'Treatment' and 'Cure.' We commend their perusal to all practical physicians."—*Dublin Journal of Medical Science*, August, 1879.

CLINICAL LECTURES ON MEDICINE AND SURGERY. Translated from the German, and selected from Professor Volkmann's Series. Two Volumes.

MEMOIRS ON DIPHTHERIA; containing Memoirs by Bretonneau, Trousseau, Daviot, Guersant, Bouchet, Empis, &c. Selected and Translated by Dr. R. H. Semple.

"Bretonneau's memoir must be considered the fullest and most searching that has yet appeared in any country on this extraordinary disease."—*British Medical Journal.*

"Like honour is due to M. Bretonneau for his admirable investigations. . . . . . His treatise on Diphtheria constitutes the greater part of the volume recently published by the New Sydenham Society. Of the remaining memoirs each contains much valuable material. . . . . . There is no part of the volume which will better repay study than the researches of M. Empis."—*Medical-Chirurgical Review.*

RADICKE'S PAPERS ON THE APPLICATION OF STATISTICS TO MEDICAL INQUIRIES. Translated by Dr. Bond.

"We can hardly conceive an object to which the New Sydenham Society could better devote a portion of its rapidly-increasing resources than to the introduction of papers such as these to the profession. It is by such work as this that the Society is calculated to confer inestimable benefits on the profession of this country."—*Medical Times and Gazette*, January 25, 1862.

LECTURES ON PHTHISIS. By Professor Niemeyer. Translated by Professor Baumler.

"Niemeyer's work is eminently suggestive, not only as regards pathology, but also as regards treatment and prevention. There is no work on treatment of Phthisis in the English language so advanced in its pathology; it leaves the crude theories of Laennec and his followers far in the rear, and by showing the essential dependence of tubercle on preceding inflammatory processes, it shows also how we may ward off this intractable disease from our patients, and how we may most usefully employ the remedies at hand for its prevention."—*Edinburgh Medical Journal*, December, 1870.

"The members of the New Sydenham Society must be well content with the works supplied to them for their subscription. Those issued of late are of peculiarly solid and lasting value. We have now three before us, which, besides the recommendation of intrinsic scientific value, have that of high practical utility. We refer to Trousseau's 'Clinical Medicine,' Niemeyer's 'Lectures on Pulmonary Consumption,' and Stricker's 'Histology.'"—*Brit. and For. Med. Chir. Rev.*, April, 1871.

THE COLLECTED WORKS OF DR. ADDISON. Edited, with Introductory Prefaces to several of the Papers, by Dr. Wilks and Dr. Daldy. With Portrait, and numerous Lithographic Plates.

"We must cordially commend the decision of the Council of the New Sydenham Society, which led to the publication of this historically interesting and practically valuable book. Few names have, of late years, been better known to the profession than that of the eminent physician whose contributions to its literature, too few in number, have nevertheless been, one and all, highly and justly esteemed. A brief but kindly and discriminating biography of Dr. Addison precedes the collection of his papers. . . . . ."—*Edinburgh Medical Journal*, December, 1868.

"No one who has studied the valuable papers, published by Dr. Addison in the Guy's Hospital Reports, can fail to be pleased that they are now rendered more widely available by this separate publication. His great and extensive knowledge of skin diseases renders the articles on that subject of much interest. If, however, we were asked to select the one most likely to be useful to the practitioner, we should unhesitatingly point to that on the Physical Examination of the Chest."—*Medical Times and Gazette*, July 4, 1868.

A GUIDE TO THE QUALITATIVE AND QUANTITATIVE ANALYSIS OF THE URINE. By Dr. C. Neubauer and Dr. J. Vogel. Fourth edition, considerably enlarged. Translated by William O. Markham, F.R.C.P.L. With four Lithographs, and numerous Woodcuts.

" The New Sydenham Society have conferred a benefit, not only on their own subscribers, but on the whole profession in this country, by publishing the work of Drs. Neubauer and Vogel."—*Medical Times and Gazette.*

" It is one of those works in which there is not an unnecessary line nor even a word. It is quite a text-book upon urinology for the scientific physician, and may be handled likewise by the youngest student."—*Lancet.*

MEMOIRS ON ABDOMINAL TUMOURS AND INTUMESCENCE. By Dr. Bright. Reprinted from the " Guy's Hospital Reports," with a Preface by Dr. Barlow. Numerous Woodcuts.

" Dr. Bright's object was to bring his vast clinical experience and great diagnostic tact to bear on the elucidation of confessedly a most obscure department of medical disease--the discrimination and diagnosis of abdominal tumours; and this he has done by briefly stating their principal characteristics, as they are produced, either by the presence of tumours dependent on a cephalocyst hydatid, by ovarian tumours, or diseases of the spleen, liver, or kidney. Under each of these heads we have valuable features recorded, by which in life they may be recognised, whilst after death their pathological characters are described in a manner that leaves but one impression on our minds, that here indeed the author has held up the mirror to nature; and under each section we have a perfect *embarras de richesse*, in the shape of illustrative cases. The whole work is profusely filled with woodcuts and outlines descriptive of the several diseases described, by which means the author's verbal descriptions are more vividly presented to the reader's understanding."—*Dublin Quarterly Journal of Medical Science*, May, 1861.

" The memoirs possess a permanent value, as models of clinical reports, as exhibiting the method by which the investigation of this difficult class of organic diseases may be pursued with greatest certainty of success, and as furnishing the great general outlines of the inquiry. It is by the study of such models that the difficult art of medical observation may best be understood, and may to some extent be acquired. Certainly no papers in our periodical literature were more worthy than these of being republished and circulated in a collected and accessible form."- *Edinburgh Medical Journal*, January, 1861.

A CLINICAL ACCOUNT OF DISEASES OF THE LIVER. By Prof. Frerichs. 2 vols. Translated by Dr. Murchison. With coloured Lithographs, and numerous Woodcuts.

" Frerichs' book is one of those treatises that will frequently be taken down from the book-shelves to be consulted, both by physiologists and physicians."—*Lancet.*

"We shall look forward with interest to the completion of this very valuable addition to the Clinical History of Liver Diseases."—*Medical Times and Gazette.*

CZERMAK ON THE PRACTICAL USES OF THE LARYNGOSCOPE. Translated by Dr. G. D. GIBB. Numerous Woodcuts.

"What has been given will, we trust, convince any one who may hitherto have doubted the value of laryngoscopy, that it is a real acquisition. To those who are desirous of becoming more fully acquainted with the subject, we strongly recommend the study of the work [Professor Czermak's] from which we have chiefly culled our extracts."—*Medico-Chirurgical Review*, Oct., 1862.

A HAND-BOOK OF PHYSICAL DIAGNOSIS COM-PRISING THE THROAT, THORAX, AND ABDOMEN. By Dr. PAUL GUTTMANN, of Berlin. Translated by Dr. NAPIER, of Glasgow.

"We are persuaded that if the practitioner will carefully study this work, and conscientiously carry out its suggestions, he will find an incalculable advance in the realistic appreciation of diseases by means of their physical phenomena. The work is not properly a 'students'' book. It presumes a certain familiarity with the diseases of the organs with which it deals, and the endeavour is made to connect the physical phenomena with the pathological conditions present in these diseases. . . . . . It was a wise decision of the New Sydenham Society to place a translation of it in the hands of their subscribers." —*Glasgow Medical Journal*, March, 1880.

"The New Sydenham Society has done well to put within the reach of their subscribers a work which not only has attained to a third edition in its own language, but has also been translated into Italian, Russian, Spanish, French, and Polish. . . . . . . As a systematic and scientific treatise it well repays perusal. The book concludes with a good account of laryngoscopy, and of the physical signs of the principal diseases of the larynx. The acoustics of percussion and auscultation are elaborated with great care, and the precise explanation of the causes of many familiar physical signs will be very acceptable to teachers of clinical medicine, who have hitherto felt the want of an adequate scientific exposition of the principles of physical diagnosis."—*Dublin Journal of Medical Science*, November, 1880.

AN ATLAS OF ILLUSTRATIONS OF PATHOLOGY, COMPILED (CHIEFLY FROM ORIGINAL SOURCES) FOR THE SOCIETY.

The Committee in charge of this work consists of Dr. GEE, Dr. GREEN, Dr. MOXON, Dr. SUTTON, Mr. HOLMES, and Mr. HUTCHINSON.

TWO FASCICULI have been published, and it is proposed to issue one every year.

The following subjects have been illustrated :—

FIRST FASCICULUS.

Scrofula; Syphilis; and Lymph-Adenoma.—Plate I.

Fig. 1. Scrofulous Disease of the Kidney and Ureter. Fig. 2. Scrofulous Disease of the Kidney. Fig. 3. Scrofulous Disease of the Kidney. Fig. 4. A Mass of Syphilitic Deposit in the Cortical Substance of the Kidney. Fig. 5. Lymph-Adenoma of Kidney.

Nephritis after Diphtheria; Scarlet Fever; and Burns.—Plate II.

Fig. 1. Nephritis after Diphtheria.—Section of Kidney. Fig. 2. Subacute Nephritis after Scarlet Fever.—Outer surface of kidney. Fig. 3. Subacute Nephritis after Scarlet Fever. Fig. 4. Acute Nephritis after Scarlet Fever. Fig. 5. Subacute Nephritis after Scarlet Fever. Fig. 6. Acute Nephritis after a Burn.—Outer surface of the kidney of a child who died after a very extensive burn. Fig. 7. Acute Nephritis after a Burn.—Section of the same kidney.

The Granular Kidney in different stages.—Plate III.

Fig. 1. Extremely Granular Kidney. Fig. 2. Extremely Granular Kidney.—Section of the same kidney. Fig. 3. Less Granular (contracted) Kidney.—Outer surface of the right kidney taken from the same subject as the left kidney shown in Figs. 1 and 2. Fig. 4. Granular Kidney of Bright. Fig. 5. Contracted Granular Kidney, in section. Fig. 6. Contracted Granular Kidney; exterior. Fig. 7. Large Granular Kidney. Fig. 8. Large Granular Kidney with cysts.

Embolism; Infarction Processes from Pyæmia; Jaundice and Purpura; Scrofula.—Plate IV.

Fig. 1. Embolic Changes in Pyæmia. Fig. 2. Embolic Changes in Pyæmia. Fig. 3. Pyæmic Deposits in Kidney. Fig. 4. Pyæmic Deposits in the Kidney. Fig. 5. Results of Jaundice and Purpura. Fig. 6. A variety of the Scrofulous Kidney.—The substance of the kidney is wholly destroyed and replaced by cavities containing a white mortar-like substance.

SECOND FASCICULUS.

Diseases of the Kidney.—Plate V.

Fig. 1. Amyloid Disease of Kidney in advanced stage. Fig. 2. A section of the same Kidney. Fig. 3. The pale flabby Kidney. Fig. 4. The same organ seen in section. Fig. 5. Medullary Cancer of the Kidney.

Various Diseased Conditions of the Spleen.—Plate VI.

Fig. 1. Hodgkin's Disease of Spleen (Lympho-sarcoma). Fig. 2. Acute Splenic enlargement in Diphtheria. Fig. 3. Suppurating infarction of Spleen from a case of Ulcerative Endocarditis. Fig. 4. Embolic changes in Pyæmia. Fig. 5. Rupture of the Spleen.

Diseases of the Supra Renal Capsules and Spleen.—Plate VII.

Fig. 1. Cancer of the Supra Renal Capsule. Figs. 2, 3, 4. Adenoma of the Supra Renal Capsule. Fig. 6. Addison's Disease of the Supra Renal Capsule (in section). Fig. 5. Addison's Disease of the Supra Renal Capsule.—"Fibro-calcareous or strumous disease." Fig. 7. Tubercle of the Spleen (external surface). Fig. 8. Tubercle of the Spleen (in section). Fig. 9. Lardaceous Spleen.

Microscopic Pathology of Kidneys.—Plate VIII.

Fig. 1. Lardaceous Degeneration of the Kidney.—Section of cortex. Fig. 2. Lardaceous Degeneration.—*g*. A glomerulus from the same kidney, as in Fig. 1, which has undergone lardaceous degeneration and is becoming fatty. Fig. 3. Part of the same seen with a higher power, showing contents of one of the tubules. Fig. 4. Lardaceous Degeneration in earlier stage combined with interstitial fibrous change. Figs. 5 & 6. Lardaceous Degeneration (after Cornil). Fig. 5. Section showing the hyaline membranous wall of the tubules *a a* much swollen, stained violet-red, showing waxy degeneration. Fig. 6. Transverse section of one of the pyramids, near summit of cone. Fig. 7. Granular Contracted Kidney. Fig. 8. From the same.—A thickened arteriole surrounded by fibroid growth. Fig. 9. Partial Fibrous Degeneration of Malpighian body in slight chronic intertubular nephritis. Fig. 10. From the same kidney; showing early changes around Malpighian body. Fig. 11. Multiplication of Nuclei on glomerulus with adhesion of capillary tuft to wall of capsule. Fig. 12. Subacute Interstitial Nephritis with large white kidney. Fig. 13. Scarlatinal Nephritis.—Intertubular exudation in a case fatal on 7th day of fever. Fig. 14. Subacute Interstitial Nephritis. Fig. 15. Acute Catarrhal Nephritis, showing swelling and granular degeneration of epithelium. (100 diam.) Fig. 16. Part of the same seen with a higher power. Fig. 17. Section of cortex from a case of parenchymatous (catarrhal) nephritis at a later stage (so-called "fatty" kidney). Fig. 18. From nearly transverse section near base of pyramid in similar case. Fig. 19. Casts in tubes in interstitial nephritis (post scarlatinal). Fig. 20. Colloid cast, *b*, in tubule; *a*, unaltered epithelium.

Microscopic Pathology of the Kidney.—Plate IX.

Fig. 1. Scarlatinal Nephritis. Fig. 2. Shows two of the glomeruli from same section as Fig. 1. Fig. 3. Section from the same.—Part of the wall of a Malpighian body from which the capillary tuft has fallen out. Fig. 4. Scarlatinal Nephritis.—(From a case fatal about 12 weeks from attack of fever). Fig. 5. Scarlatinal Nephritis.—(From a case fatal 15 months after attack of scarlet fever). Fig. 6. From same kidney as Fig. 5, but in a deeper part of cortex, close to medulla. Similar growth of interstitial connective tissue. Fig. 7. Subacute Interstitial Nephritis, probably Scarlatinal, under low power; showing diffuse infiltration and cluster of dilated tubules. Fig. 8. Chronic Parenchymatous Nephritis (large white kidney) with little or no interstitial change. Section of cortex, showing changes in epithelium of convoluted tubules. Fig. 9. Kidney in leucocythæmia—to show localisation of changes around glomeruli and vessels. Fig. 10. Swelling of inner coat of small artery in granular contracted kidney. Fig. 11. Tuberculous Pyelonephritis. Fig. 12. Fatty Degeneration from Alcoholic Poisoning (after Lancereaux). Fig. 13. Fatty Degeneration in Cancer. Fig. 14. Individual epithelial cells from the preceding section; in various stages of fatty degeneration. Fig. 15. Cystic Degeneration of Kidney (after Lancereaux.) Fig. 16. From a cyst in kidney near base of pyramid. Fig. 17. Colloid Degeneration of Kidney. Figs. 18, 19, 20, and 21, illustrate the hyaline changes found in the splenic arteries in certain febrile conditions. Fig. 18. From a section through the spleen of a case of early scarlatina, showing hyaline degeneration of the coat of an artery, transversely cut. Fig. 19. Artery in longitudinal section. Fig. 20. Malpighian corpuscle from the spleen of a case of early scarlatina. Fig. 21. Part of the central and intermediate zone of the same Malpighian corpuscle as in Fig. 20, only more highly magnified (180 diam.) Fig. 22. Hodgkin's Disease. — Section of a spleen to show the overgrowth of the lymphatic sheath in Hodgkin's disease. (1 inch.) Fig. 23. Adenoma of the Supra Renal Capsule, showing the columns stuffed with fatty granules.

Microscopic Pathology of Spleen and Supra-renals.—Plate X.

Fig. 1. Capsulitis of the Spleen.—Vertical section of fibrous nodule in the capsule of the spleen, showing that the thickening of the capsule takes place by cellular growth in its deeper layers. Fig. 2. Fibrosis of the Spleen.—From the enlarged spleen of a ricketty child. Fig. 3. Fibrosis of the Spleen.—Showing a more advanced or fibrous condition spreading round some dilated veins. Fig. 4. Muscular Hypertrophy.—Over-growth of muscular trabeculæ in the spleen. Fig. 5. Muscular Hypertrophy. — Extreme stage of fibro-muscular growth in the spleen. Fig. 6. The Leucocythæmic Spleen.—Section of the edge of a Malpighian corpuscle, showing the compressed fibrous tissue between it and the splenic pulp. Fig. 7. The Leucocythæmic Spleen. — The pulp and stroma are normal. Fig. 8. Hodgkin's Disease.—The texture of a lymphoid noduie in the spleen of Hodgkin's disease. Fig. 9. Tubercular Spleen. (37 diam.) Fig. 10. Tubercular Spleen. Fig. 11. Induration and Atrophy. — A section of the spleen from a case of heart disease. Fig. 12. Lardaceous Spleen.—The sago spleen, showing the Malpighian corpuscles and small arteries mapped out by structureless hyaline lardaceous matter. Fig. 13. Lardaceous Spleen.—Transverse section of a Malpighian corpuscle, or small artery, with its surrounding lymphoid sheath. Fig. 14. Addison's Disease.—Vertical section of a supra renal capsule from the exterior inwards, to show the early changes in Morbus Addisonii. (250 diam.) Fig. 15. Addison's Disease.—Section of a supra renal capsule, to show the late, or fibro-calcareous, stage of Morbus Addisonii.

With Essay on the Pathology of the Kidney, by Dr. Greenfield. Essay on the Pathology of the Spleen and Supra-renals, by Dr. Goodhart.

In Preparation.

ON THE DISEASES OF OLD AGE. By Prof. CHARCOT. Translated by Mr. WILLIAM TUKE.

In Preparation.

THE DIAGNOSIS AND TREATMENT OF DISEASES OF THE CHEST. By Dr. STOKES. A Reprint to be Edited by Dr. HUDSON, of Dublin.

Surgery.

ESMARCH ON THE USES OF COLD IN SURGICAL PRACTICE. Translated by Dr. MONTGOMERY. Woodcuts.

"Esmarch's treatise is of high practical interest."—*British Medical Journal*, December, 1863.

BILLROTH'S LECTURES ON SURGICAL PATH-OLOGY AND THERAPEUTICS. A Hand-book for Students and Practitioners. 2 vols.

"While being rendered in most fluent and unconstrained English, it is singularly free from obscurities and ambiguities with which translations generally abound."—*London Medical Record*, April, 1878.

"Whether looked at as a text-book for students or as a work of reference for the hard-worked and busy practitioner, it deserves to be spoken of in high terms of commendation."—*Brit. and For. Med. Chir. Rev.*, July, 1873.

INVESTIGATION INTO THE ETIOLOGY OF THE TRAUMATIC INFECTIVE DISEASES. By R. Koch. Translated, with Lithographic Plates, by Mr. Watson Cheyne.

ON THE PROCESS OF REPAIR AFTER RESECTION AND EXTIRPATION OF BONES. By Dr. A. Wagner, of Berlin. Translated by Mr. T. Holmes.

CLINICAL LECTURES. Selected from Professor Volkmann's Series. 2 vols. (See "Medicine.")

In Preparation.

THE WORKS OF ABRAHAM COLLES. Chiefly his Treatise on the Venereal Disease and on the Use of Mercury. Edited, with Portrait, by Dr. McDonnell, of Dublin.

Gynaecology.

ON THE MORE IMPORTANT DISEASES OF WOMEN AND CHILDREN, with other Papers, by Dr. Gooch. Reprinted; with a Prefatory Essay by Dr. Robert Ferguson. With woodcuts.

"The work of Dr. Gooch is so well known and highly appreciated by every lover of medical literature that we need say nothing in its praise. It has been before the world for thirty years, and only one opinion has been expressed upon its merits. We cannot but consider, therefore, that the Council of the New Sydenham Society has done well to republish it, more especially as the Council has had the good fortune to persuade Dr. Robert Ferguson to furnish an introductory essay on the author's life and writings."—*Lancet.*

CLINICAL MEMOIRS ON DISEASES OF WOMEN. By Drs. Bernutz and Goupil. 2 vols. Translated and abridged, Dr. Meadows.

"The careful study of these valuable memoirs is imperative on all who are interested in gynaecology."—*Lancet*, October, 1866.

SMELLIE'S MIDWIFERY. 3 vols. Edited and Annotated by Dr. McClintock, of Dublin. With Portrait of Smellie.

"This book begins with a fine engraving of the author, and had the N. S. S. done for Smellie's memory no more than the publication of this valuable print, it would have a strong claim on the gratitude of the profession. McClintock's life of Smellie is a very interesting contribution to medical literature. His works show that he was a very great man and midwife, but his biography was needed to show his peculiarities. . . . . . Let the reader carefully peruse Dr. McClintock's annotations, and he will see how Smellie's Editor recognises Smellie's keenness of eye in discerning how to make progress."—*Edin. Med. Journal*, March, 1877.

"The New Syd. Soc. has done nothing more commendable than to produce the work we are now about to notice. . . . . . Smellie was the Sydenham of Midwifery. Although it was a chief part of his glory to have studied deeply and soundly the mechanism of labour as a natural process, and in that study to have laid the ample foundations of the highly finished art of midwifery as we see it practised by the best obstetricians of the present day, we also see evidence in every one of his 'cases' of shrewd and sagacious medical views, showing that his great manipulative faculties were governed and controlled by good judgment, physiological considerations, and that great respect for nature which is a characteristic of all great physicians. . . . . . In short, he was a model practitioner in midwifery whose influence grows rather than diminishes, and whose works will be found to contain the germ of most of our practice and doctrine. . . . . . Dr. McClintock has fairly placed alongside of Smellie's principal views those of modern authorities, including his own, derived from an experience altogether exceptional, and has produced a joint work without which no obstetric library will be complete."—*Lancet*, August 4, 1877.

Diseases of the Eye and Ear.

ON THE ANOMALIES OF ACCOMMODATION AND REFRACTION OF THE EYE, with a PRELIMINARY ESSAY ON PHYSIOLOGICAL DIOPTRICS. By F. C. DONDERS, M.D., Professor of Physiology and Ophthalmology in the University of Utrecht. Written expressly for the Society. Translated from the Author's Manuscript by W. D. MOORE, M.D.

"This splendid monograph, from the hand of the accomplished professor of physiology and ophthalmology, of Utrecht, will be hailed as a boon by all lovers of ophthalmic science."—*Lancet*.

THREE MEMOIRS ON GLAUCOMA AND ON IRIDECTOMY AS A MEANS OF TREATMENT. By Professor VON GRÆFE. Translated by Mr. T. WINDSOR, of Manchester.

"This is the fifth volume of the first year, and contains translations of three mportant and well known essays from the German."—*Lancet*.

"The value—the great practical value—of these memoirs will be admitted by every one who peruses them."—*Medical Times and Gazette*.

ON THE MECHANISM OF THE BONES OF THE EAR AND THE MEMBRANA TYMPANI. (Pamphlet.) By Professor HELMHOLTZ. Translated by Mr. HINTON.

"This little work is the translation of a very valuable essay published by the great physicist of Berlin, and which is thus rendered accessible to a wide circle of English readers."—*Lancet*, July 5, 1873.

THE AURAL SURGERY OF THE PRESENT DAY. By W. KRAMER, M.D., of Berlin. Translated by HENRY POWER, Esq., F.R.C.S., M.B. With two Tables and nine Woodcuts.

VON TROELTSCH'S TREATISE ON DISEASES OF THE EAR. Translated, with Notes, by Mr. HINTON.

Forensic Medicine.

A HANDBOOK OF THE PRACTICE OF FORENSIC MEDICINE, BASED UPON PERSONAL EXPERIENCE. By J. L. CASPER, M.D., late Professor of Medical Jurisprudence in the University of Berlin. Translated by G. W. BALFOUR, M.D. 4 vols.

"Casper's great work, based as it is upon a minute and laborious observation of facts, must prove the most trustworthy guide in the interpretation of the ofttimes difficult questions which the medical jurist is called upon to solve."—*Lancet.*

"This work must be regarded as a valuable and judicious addition to the publications of the Society from which it emanates. The advantages to be derived by the reader from its perusal cannot be over-estimated or too eagerly sought for."—*Madras Quarterly Journal of Medical Science.*

Diseases of the Nervous System.

SCHRŒDER VAN DER KOLK ON A CASE OF ATROPHY OF THE LEFT HEMISPHERE OF THE BRAIN. Translated by Dr. W. MOORE, of Dublin. Four Lithographs.

ON THROMBOSIS OF THE CEREBRAL SINUSES. By Professor VON DUSCH. Translated by Dr. WHITLEY.

LECTURES ON DISEASES OF THE NERVOUS SYSTEM. By Professor CHARCOT. (First Series.) Translated by Dr. SIGERSON, of Dublin. With woodcuts.

"These lectures of M. Charcot are too well known in the original to call for any special criticism here. They have, indeed, obtained an European reputation, and it has long been felt that it would be a great gain to our literature to have them rendered into English. . . . . We strongly advise all those of our readers who may not yet have made themselves acquainted with these lectures to lose no time in doing so. . . . . The translator, Dr. Sigerson, a former pupil of the author, has succeeded admirably in his rendering of the elegant literary style of M. Charcot. . . . . It is, without doubt, one of the most valuable books that has been issued by this Society since their translation of Trousseau."—*Lancet*, August, 1877.

"This volume will be highly prized by the members of the N. S. S. M. Charcot's name ranks among the very foremost of those who have advanced the knowledge of nerve-pathology. The work he has done is marked by great accuracy and close observation, and by great acumen in interpreting facts and drawing inferences."—*Brit. and For. Med. Chir. Rev.*, July, 1877.

In Preparation.

A SECOND VOLUME OF LECTURES ON DISEASES OF THE NERVOUS SYSTEM. By Professor CHARCOT. Translated by Dr. SIGERSON. With this volume all the Plates to the two volumes will be given.

A MANUAL OF MENTAL PATHOLOGY AND THERAPEUTICS. By Professor GRIESINGER. Translated by Dr. LOCKHART ROBERTSON and Dr. JAMES RUTHERFORD.

"The thanks of the profession are due to the Council of the N. S. S. for the selection of this work. . . . . . We need scarcely say that each section is full of instruction, and carries upon its face the evidence of great experience and close and deep thought."—*Medical Times and Gazette*, September, 1867.

ON EPILEPSY. By Professor SCHRŒDER VAN DER KOLK.

Anatomy, Physiology, and General Pathology.

A MANUAL OF HUMAN AND COMPARATIVE HISTOLOGY. By S. STRICKER. 3 vols. Translated by Mr. POWER.

"This work, edited by Stricker, and having as its contributors nearly all of the best names in Germany, is one well deserving of attention, and constitutes, we think, a very valuable addition to the stores of the New Sydenham Society." —*Medical Times and Gazette*, December 10, 1870.

"There has hitherto been no work which contained a full and complete account of the various elements of animal structure, still less of the way in which minute examination of these elements should be conducted. The book before us supplies this want in a very remarkable degree. . . . . The work is illustrated by over a hundred woodcuts. Modern medical literature of the higher class so teems with histological references, that a treatise in which they are explained has become almost a necessity."—*Lancet*, December 3, 1870.

"We must congratulate the New Sydenham Society on their enterprise, and thank them for making so important a work accessible to the English reader."—*Quarterly Journal of the Microscopic Society*, April, 1873.

"Ably translated and edited by Mr. Henry Power. . . . . The members of the Society may be congratulated on the addition of such valuable treatises to their libraries."—*Brit. and For. Med. Chir. Rev.*, July, 1873.

EXPERIMENTAL RESEARCHES ON THE EFFECTS OF LOSS OF BLOOD IN PRODUCING CONVULSIONS, By Drs. KUSSMAUL and TENNER. Translated by Dr. BRONNER, of Bradford.

A MANUAL OF PATHOLOGICAL HISTOLOGY, intended to serve as an introduction to the study of Morbid Anatomy. By Professor RINDFLEISCH. (Bonn.) 2 vols. Translated by Dr. BAXTER.

VOL. 1.—"Rindfleisch's work forms a mine which no recent pathological writer could afford to neglect who desired to interpret aright pathological structural changes. . . . . . The special part treats of the anomalies of the blood, the circulatory apparatus, of the serous and mucous membranes, skin, lung, liver, kidneys, and so on. As a specimen of the scientific spirit with which Rindfleisch has entered upon his very laborious work, the reader cannot do better than to peruse the part devoted to normal as a type of the pathological growths, and that which immediately follows on interstitial inflammation and specific inflammation. . . . . . Altogether the book is the result of honest, hard labour."—*Lancet*, April 6, 1872.

VOL. 2.—"The members of the Society may be congratulated on the addition of such valuable treatises to their libraries. . . . . . The Society ought to flourish whilst it caters so well for its members. They have every reason to be content both with the quantity and quality of the matter supplied."—*Brit. and For. Chir. Rev.*, July, 1873.

AN ATLAS OF ILLUSTRATIONS OF PATHOLOGY. (See "Medicine," page 11.)

ON THE MINUTE STRUCTURE AND FUNCTIONS OF THE SPINAL CORD. By Professor SCHRŒDER VAN DER KOLK. Translated by Dr. W. D. MOORE. Numerous Lithographs.

ON THE MINUTE STRUCTURE AND FUNCTIONS OF THE MEDULLA OBLONGATA, AND ON EPILEPSY. By Professor SCHRŒDER VAN DER KOLK. Translated by Dr. W. D. MOORE. Numerous Lithographs.

Retrospects, and Works of General Reference.

A YEAR-BOOK OF MEDICINE AND SURGERY, AND THEIR ALLIED SCIENCES, for 1859. Edited by Dr. HARLEY, Dr. HANDFIELD JONES, Mr. HULKE, Dr. GRAILY HEWITT, and Dr. ODLING.

"Our space will not admit of a further statement of the excellent character of the Year-Book and the other works issued by the New Sydenham Society, but we should strongly urge every member of the profession, who has the advancement of medical knowledge at heart, to lose no time in forwarding his name, should he not already have done so."—*London Medical Journal.*

YEAR-BOOK for 1860. Edited by Dr. Harley, Dr. Handfield Jones, Mr. Hulke, Dr. Graily Hewitt, and Dr. Sanderson.

"This is, as it professes to be, an improvement on its predecessor. On the whole the editors have done their laborious work well."—*British Medical Journal*, December 31, 1861.

YEAR-BOOK for 1861. Edited by Dr. Harley, Dr. Handfield Jones, Mr. Hulke, Dr. Graily Hewitt, and Dr. Sanderson.

YEAR BOOK for 1862. Edited by Dr. Montgomery, Dr. Handfield Jones, Mr. Windsor, Dr. Graily Hewitt, and Dr. Sanderson.

YEAR-BOOK for 1863. By the same Editors.

YEAR-BOOK for 1864. Edited by Mr. Hinton, Dr. Handfield Jones, Mr. Windsor, Dr. M. Bright, and Dr. Hilton Fagge.

"Of the usefulness of these reports all who have consulted them will bea the fullest testimony. They supply a very valuable bibliography; they enable the reader to judge what papers or works he may study with advantage to his peculiar pursuits; and they present a condensed summary of the most important advances and improvements in medical science."—*Edinburgh Medical Journal.*

A BIENNIAL RETROSPECT OF MEDICINE, SURGERY, AND THEIR ALLIED SCIENCES, for the Years 1865 and 1866. Edited by Mr. Power, Dr. Anstie, Mr. Holmes, Dr. Barnes, Mr. Windsor, and Dr. Hilton Fagge.

A BIENNIAL RETROSPECT OF MEDICINE, SURGERY, AND THEIR ALLIED SCIENCES, for the Years 1867 and 1868. Edited by Mr. H. Power, Dr. Anstie, Mr. Holmes, Mr. R. B. Carter, Dr. Barnes, and Dr. Thomas Stevenson.

A BIENNIAL RETROSPECT for 1869 and 1870.

"As to the Biennial Retrospect, it is as good as any of its class; while of little value to town practitioners, possessing easy access to large, well-selected, and well-catalogued libraries, it is no doubt of great value to country practitioners whose resources in that respect are more limited." — *Edinburgh Medical Journal*, May, 1872.

A BIENNIAL RETROSPECT for 1871 and 1872.

A BIENNIAL RETROSPECT for 1873 and 1874.

"Full justice is done to English observers, and the whole volume is creditable to its compilers and to the Society under whose auspices it is published."—*Lancet*, January, 1876.

THE MEDICAL DIGEST. Being a means of ready reference to the principal contributions to Medical Science during the last Thirty years. By Dr. Richard Neale.

"The Council has certainly acted wisely in publishing the work before us It is a section of what has long been a desideratum—a general index to medical literature, and as a section its great value cannot but suggest how inestimably valuable a complete work of this kind would be. . . . . . Compiled by a practitioner for his own use, it is calculated especially for the use of the practitioner."—*Lancet*, January 5, 1875.

"The idea of this volume is a good one. Something of the kind had been all along contemplated by the Society, but never carried out till now, when Dr. Neale offered his manuscript, exactly as it is printed. . . . . . We have been at the pains of testing the index in a good many instances, and have come to the conclusion that it may be relied on for discovering easily the contents of the volume."—*Edinburgh Medical Journal*, April, 1878.

BIBLIOTHECA THERAPEUTICA; OR BIBLIOGRAPHY OF THERAPEUTICS. By E. J. Waring, M.D. 2 vols.

"We feel sure that, although not exactly what we would like in a work of the kind, Waring's 'Bibliotheca Therapeutica,' with its copious and valuable indices, will be frequently referred to with advantage, and with considerable confidence as regards its accuracy."—*Glasgow Medical Journal*, Sept., 1879.

"With the Index of Diseases before him, the student has a bird's-eye view of the principal remedies recommended from time to time in the treatment of individual diseases, and the dates of their respective employments; whilst further reference to the body of the work, in the manner pointed out in the index, will disclose the name of the authority, and other particulars of special interest to the pathologist and therapeutist."—From *Preface* to Vol. 2.

A LEXICON OF MEDICAL TERMS. Edited by Mr. Power and Dr. Sedgwick. Parts 1 to 4. This Lexicon is based upon the well-known work of Dr. Mayne, the copyright of which was purchased by the Society. It is, however, entirely rewritten by the present Editors, and very much enlarged.

Diseases of the Skin and Syphilis.

ON SYPHILIS IN INFANTS. By Paul Diday. Translated by Dr. Whitley.

"The work of M. Diday is of great merit; it contains all that has been written on infantile syphilis, and he puts the whole subject in a well-arranged form for further investigation as well as present use."—*Brit. and For. Med. Chir. Rev.*

ON DISEASES OF THE SKIN, INCLUDING THE EXANTHEMATA. By Professor Hebra. 5 vols. Translated and Edited by Dr. Hilton Fagge, Dr. Pye-Smith, and Mr. Waren Tay.

"Had we space we should have been glad to enter into a lengthened critique of the second volume of Hebra's work. We are relieved from any misgiving, however, by the fact that the work will be very largely circulated amongst our readers by the Sydenham Society, and that they, with others who aspire to any real knowledge of skin diseases, would not, under any circumstances, be satisfied without studying the work for themselves. . . . . . This second volume contains information relative to the most important diseases of the skin; and it will, we are confident, do good service in helping on the cause of cutaneous medicine in England."—*Lancet*, November 7, 1868.

"Of all the works produced by the New Sydenham Society this is one of the most valuable and most welcome. . . . . . It is to be remarked that this book is not a mere translation of the German work; it is a new and revised edition, undertaken by the author for his English brethren."—*Medical Times and Gazette*, April 27, 1867.

"The New Sydenham Society has done good service to the medical profession by undertaking the translation and publication of Professor Hebra's excellent work. In several respects the English edition is greatly superior to the original. In closing its pages we have but one regret, namely, that the New Sydenham Society does not embody the whole medical confraternity, so that every member af our noble profession might have on his bookshelves a copy of this most valuable book."—*Journal of Cutaneous Medicine*, April, 1877.

VOL. 3.—"Mr. Tay has performed a difficult task with great ability and success, and the work is far pleasanter to read in its English dress than in the original. Mr. Tay has enriched the work with valuable notes of his own, embodying the views of English authorities and sometimes his own experience on the question discussed in the body of the work."—*Medical Times and Gazette*, June 20, 1874.

VOL. 4—"The entire work is admirable for its lucidity of arrangement, its simplification of confused and intricate subjects, and not least for the avoidance of those pedantic and repelling terms which a celebrated dermatologist has grandiloquently styled the 'terminological innovations of modern nomenclators.'"—*Dublin Journal of Medical Science*, May, 1875.

LANCEREAUX'S TREATISE ON SYPHILIS. 2 vols. Translated by Dr. WHITLEY.

"The work is the most exhaustive book which has been published on the subject, and has been quoted by all the recent writers in this country, America, and the Continent. It is a perfect mine of information. The translation is well done, and the New Syd. Soc. may be congratulated on having added such an important treatise to its list of works."—*Lancet*, March, 1869.

The Society's Atlas of Diseases of the Skin.

In fifteen Annual Fasciculi comprising the following subjects. Unless otherwise indicated, the Plates are original.

	PLATE
Favus. From Hebra	I.
Tinea Tonsurans. From Hebra.	II.
Lupus Exulcerans. From Hebra.	III.
Psoriasis Diffusa. From Hebra.	IV.
Ichthyosis. From Hebra.	V.
Lupus Serpiginosus; Alopecia Areata. From Hebra. .	VI.

		PLATE
Framboesia (Endemic Verrugas)	. . . .	XLI.
Lupus Erythematosus	. . . .	XLII.
Ulcerating Eruption from Bromide of Potassium	. . .	XLIII
Morphœa, or Addison's Keloid	. . . .	XLIV.

"This Fasciculus supplies life size portraits of pityriasis rubra, papular syphilis, with indurated chancres, and pruriginous impetigo following varicella, which are extremely beautiful, and look life-like."—*Edin. Medical Journal*, May, 1872.

"They are better, to our mind, than any other plates in use amongst us; and there cannot be a question as to the Society's issue being as popular as it is useful."—*Lancet.*

"We have received the thirteenth fasciculus of this splendid collection of drawings, of which no further praise is needed than to say that they are executed with the same artistic skill and fidelity to nature which have characterised the whole series."—*Dublin Journal of Medical Science*, May, 1874.

A CATALOGUE OF THE PORTRAITS COMPRISED IN THE SOCIETY'S ATLAS OF SKIN DISEASES. Prepared, at the request of the Council, by Mr. HUTCHINSON. Parts 1 and 2.

"The descriptions, cases, and plates are well given. . . . . . There is one good feature in some of the cases described. Take that of Addison's Keloid, p. 160. In it we have notes, &c., of a rare skin disease, which has been accurately described by the observers under whose care the patient had been at various stages of the case. This is, therefore, a valuable contribution to medicine."—*Edinburgh Medical Journal*, February, 1877.

LIST OF PUBLISHED WORKS.

Arranged according to the Year of Issue.

Vol. 1859. (*First Year.*)

1. Diday on Infantile Syphilis.
2. Gooch on Diseases of Women.
3. Memoirs on Diphtheria.
4. Van der Kolk on the Spinal Cord, &c.
5. Monographs (Kussmaul and Tenner, Graefe, Wagner, &c.)

1860. (*Second Year.*)

6. Dr. Bright on Abdominal Tumours.
7. Frerichs on Diseases of the Liver. Vol. I.
8. A Yearbook for 1859.
9. Atlas of Portraits of Skin Diseases. (1st Fasciculus.)

1861. (*Third Year.*)

10. A Yearbook for 1860.
11. Monographs (Czermak, Dusch, Radicke, &c.)
12. Casper's Forensic Medicine. Vol. I.
14. Atlas of Portraits of Skin Diseases. (2nd Fasciculus.)

1862. (*Fourth Year.*)

13. Frerichs on Diseases of the Liver. Vol. II.
15. A Yearbook for 1861.
16. Casper's Forensic Medicine. Vol. II.
17. Atlas of Portraits of Skin Diseases (3rd Fasciculus.)

1863. (*Fifth Year.*)

18. Kramer on Diseases of the Ear.
19. A Yearbook for 1862.
20. Neubauer and Vogel on the Urine.

VOL. 1864. (*Sixth Year.*)

21. CASPER's Forensic Medicine. Vol. III.
22. DONDERS on the Accommodation and Refraction of the Eye.
23. A YEARBOOK for 1863.
24. ATLAS of Portraits of Skin Diseases. (4th Fasciculus).

1865. (*Seventh Year.*)

25. A YEARBOOK for 1864.
26. CASPER's Forensic Medicine. Vol. IV.
27. ATLAS of Portraits of Skin Diseases. (5th Fasciculus).

1866. (*Eighth Year.*)

28. BERNUTZ and GOUPIL on the Diseases of Women. Vol. I.
29. ATLAS of Portraits of Skin Diseases. (6th Fasciculus.)
30. HEBRA on Diseases of the Skin. Vol. I.
31. BERNUTZ and GOUPIL on Diseases of Women. Vol. II.

1867. (*Ninth Year.*)

32. BIENNIAL Retrospect of Medicine and Surgery.
33. GRIESINGER on Mental Pathology and Therapeutics.
34. ATLAS of Portraits of Skin Diseases. (7th Fasciculus).
35. TROUSSEAU's Clinical Medicine. Vol. I.

1868. (*Tenth Year.*)

36. THE Collected Works of Dr. Addison.
37. HEBRA on Skin Diseases. Vol. II.
38. LANCEREAUX's Treatise on Syphilis. Vol. I.
39. ATLAS of Portraits of Skin Diseases. (8th Fasciculus).
40. CATALOGUE of Atlas of Skin Diseases. (First Part.)

1869. (*Eleventh Year.*)

41. LANCEREAUX's Treatise on Syphilis. Vol. II.
42. TROUSSEAU's Clinical Medicine. Vol. II.
43. BIENNIAL Retrospect of Medicine and Surgery.
44. ATLAS of Portraits of Skin Diseases. (9th Fasciculus.)

1870. (*Twelfth Year.*)

45. TROUSSEAU's Lectures on Clinical Medicine. Vol. III.
46. NIEMEYER's Lectures on Pulmonary Consumption.
47. STRICKER's Manual of Histology. Vol. I.
48. ATLAS of Portraits of Skin Diseases. (10th Fasciculus).

VOL. 1871. (*Thirteenth Year.*)

49. WUNDERLICH's Medical Thermometry.
50. BIENNIAL Retrospect of Medicine and Surgery.
51. TROUSSEAU's Clinical Medicine. Vol. IV.
52. ATLAS of Portraits of Skin Diseases. (11th Fasciculus.)

1872. (*Fourteenth Year.*)

53. STRICKER's Manual of Histology. Vol. II.
54. RINDFLEISCH's Pathological Histology. Vol. I.
55. TROUSSEAU's Clinical Medicine. Vol. V.
56. ATLAS of Portraits of Skin Diseases. (12th Fasciculus.)

1873. (*Fifteenth Year.*)

57. STRICKER's Manual of Histology. Vol. III.
58. RINDFLEISCH's Pathological Histology. Vol. II.
59. BIENNIAL Retrospect of Medicine and Surgery.
60. ATLAS of Portraits of Skin Diseases. (13th Fasciculus.)

1874. (*Sixteenth Year.*)

61. HEBRA on Skin Diseases. Vol. III.
62. Von TROELTSCH on Diseases of the Ear.
HELMHOLTZ on Membrana Tympani, &c. (In one Vol.)
64. ATLAS of Portraits of Skin Diseases. (14th Fasciculus.)
63. HEBRA on Skin Diseases. Vol. I.

1875. (*Seventeenth Year.*)

65. BIENNIAL Retrospect of Medicine and Surgery.
66. CATALOGUE of Atlas of Skin Diseases. (Second Part.)
67. LATHAM's Works. Vol. I.
69. ATLAS of Portraits of Skin Diseases. (15th Fasciculus.)
70. CLINICAL Lectures by various German Professors.

1876. (*Eighteenth Year.*)

68. SMELLIE's Midwifery, by McClintock.
71. CLINICAL Lectures by various German Professors.
72. CHARCOT's Clinical Lectures on Diseases of the Nervous System.
73. BILLROTH's Lectures on Surgery.

1877. (*Nineteenth Year.*)

74. SMELLIE's Midwifery, by McClintock.
75. CLINICAL Lectures by various German Professors.
76. THE Medical Digest, by Dr. Neale.
77. BILLROTH's Lectures on Surgery. Vol. II.

VOL. 1878. (*Twentieth Year.*)

79. SMELLIE'S Midwifery, by McClintock. (Concluding Volume.)
78. BIBLIOTHECA Therapeutica, by Dr. Waring. Vol. I.
80. LATHAM'S Works. Vol. II.
81. LEXICON of Medical Terms. (First Part.)

1879. (*Twenty-first Year.*)

82. BIBLIOTHECA Therapeutica, by Dr. Waring. Vol. II.
83. LEXICON of Medical Terms. (Second Part.)
84. MANUAL of Physical Diagnosis, by Dr. Guttmann.
85. ATLAS of Illustrations of Pathology. (Fasciculus II.)

1880. *Twenty-second Year.*

86. LEXICON of Medical Terms. (Third Part.)
87. HEBRA on Diseases of the Skin. Vol. V.
88. KOCH'S Researches on Wound Infection.
89. LEXICON of Medical Terms. (Fourth Part.)
90. ATLAS of Illustrations of Pathology. (Fasciculus III.)

LIST OF SURPLUS VOLUMES,

With Prices.

N.B.—The prices affixed can be continued only for a limited period until surplus stock is disposed of.

ATLAS OF SKIN DISEASES. Fasciculi I. to XV. Separately, 10s. 6d. each. Most of the stones have been destroyed, and only a limited number of impressions remain in stock, and a few are out of print.

ON SYPHILIS IN INFANTS. By PAUL DIDAY. Translated by Dr. WHITLEY. 2s. 6d.

GOOCH ON THE MORE IMPORTANT DISEASES OF WOMEN AND CHILDREN. Prefatory Essay by Dr. ROBERT FERGUSON. Woodcuts. 2s. 6d.

MEMOIRS ON DIPHTHERIA. By BRETONNEAU, TROUSSEAU, DAVIOT, GUERSANT, BOUCHUT, EMPIS, &c. Selected and Translated by Dr. R. H. SEMPLE. 3s. 6d.

ON THE MINUTE STRUCTURE AND FUNCTIONS OF THE SPINAL CORD. By Professor SCHROEDER VAN DER KOLK.

ON THE MINUTE STRUCTURE AND FUNCTIONS OF THE MEDULLA OBLONGATA, AND ON EPILEPSY. By Professor SCHROEDER VAN DER KOLK. Translated by Dr. W. D. MOORE, of Dublin. In one volume, with numerous Lithographs. 5s.

EXPERIMENTAL RESEARCHES ON THE EFFECTS OF THE LOSS OF BLOOD IN INDUCING CONVULSIONS. By Drs. KUSSMAUL and TENNER. Translated by Dr. BRONNER, of Bradford.

ON THE PROCESS OF REPAIR AFTER RESECTION AND EXTIRPATION OF BONES. By Dr. A. WAGNER, of Berlin. Translated by Mr. T. HOLMES. Numerous Woodcuts.

PROFESSOR VON GRAEFE'S THREE MEMOIRS ON GLAUCOMA, AND ON IRIDECTOMY AS A MEANS OF TREATMENT. Translated by Mr. T. WINDSOR, of Manchester.

Three Monographs in one Volume. 2s. 6d.

MEMOIRS ON ABDOMINAL TUMOURS AND INTUMESCENCE. By Dr. BRIGHT. Reprinted from the 'Guy's Hospital Reports,' with a Preface by Dr. BARLOW. Numerous Woodcuts. 7s. 6d.

A CLINICAL ACCOUNT OF DISEASES OF THE LIVER. By Professor FRERICHS. Translated by Dr. MURCHISON. Numerous Woodcuts and coloured Lithographs. 2 vols. 12s. 6d. Vol I. separately, 3s. 6d.

CZERMAK ON THE PRACTICAL USES OF THE LARYNGOSCOPE. Translated by Dr. G. D. GIBB. Numerous Woodcuts.

DUSCH ON THROMBOSIS OF THE CEREBRAL SINUSES. Translated by Dr. WHITLEY.

SCHROEDER VAN DER KOLK ON ATROPHY OF THE BRAIN. Translated by Dr. W. D. MOORE, of Dublin. Four Lithographs.

RADICKE'S PAPERS ON THE APPLICATION OF STATISTICS TO MEDICAL ENQUIRIES. Translated by Dr. BOND.

ESMARCH ON THE USES OF COLD IN SURGICAL PRACTICE. Translated by Dr. MONTGOMERY.

Five Monographs in one Volume. 5s.

A HAND-BOOK OF THE PRACTICE OF FORENSIC MEDICINE, BASED UPON PERSONAL EXPERIENCE By J. L. CASPER, M.D., Professor of Forensic Medicine in the University of Berlin. Translated by Dr. G. W. BALFOUR. Vols. II., III., IV. 7s. 6d. each. Vol. I. can be had only by subscribing for the year.

THE AURAL SURGERY OF THE PRESENT DAY. By W. KRAMER, M.D., of Berlin. Translated by HENRY POWER, F.R.C.S., M.B. With two Tables and nine Woodcuts. 2s. 6d.

A GUIDE TO THE QUALITATIVE AND QUANTITATIVE ANALYSIS OF THE URINE. By Dr. C. NEUBAUER and Dr. J. VOGEL. Fourth edition, considerably enlarged. Translated by by W. O. MARKHAM, F.R.C.P.L. With Four Lithographs and numerous Woodcuts. 5s.

ON THE ANOMALIES OF ACCOMMODATION AND REFRACTION OF THE EYE, WITH A PRELIMINARY ESSAY ON PHYSIOLOGICAL DIOPTRICS. By F. C. DONDERS, M.D., Professor of Physiology and Ophthalmology in the University of Utrecht. Translated from the Authors's Manuscript by W. D. MOORE, M.D. 7s. 6d.

TROUSSEAU'S CLINICAL MEDICINE. Vols. I., II., III., IV., and V. 5s. each, or 21s. the set. This set comprises the complete work, with copious Index.

YEAR-BOOKS OF MEDICINE AND SURGERY. 1859—66. Seven Vols. 2s. 6d. each vol.

BIENNIAL RETROSPECT OF MEDICINE AND SURGERY, 1866—75. 5 vols. 2s. 6d. each.

STRICKER'S MANUAL OF HISTOLOGY. 3 vols. 31s. 6d. Vols. I. and III. separately, 5s. each.

RINDFLEISCH'S PATHOLOGICAL HISTOLOGY. Vol. II., 5s. Vol. I. can be obtained only by subscribing for the year of its issue.

HEBRA'S TREATISE ON DISEASES OF THE SKIN. Vols. 1 to 4, 21s.

LANCEREAUX'S TREATISE ON SYPHILIS. Two vols. 5s.

NIEMEYER'S LECTURES ON PULMONARY CONSUMPTION. 2s. 6d.

LATHAM'S WORKS. 2 vols. 7s. 6d. Vol. I., 2s. 6d.

CLINICAL LECTURES BY VARIOUS GERMAN PROFESSORS. First series, 5s.

Several of these works are well suited for presents to Students or for Class Prizes. Amongst them TROUSSEAU'S Clinical Medicine; STRICKER'S Histology; FRERICH'S On Diseases of the Liver; LATHAM'S Works; and DONDERS On Anomalies of Refraction, &c., may be especially mentioned.

None of the Works published in 1876, 1877, 1878, or 1879 can be obtained separately unless by special arrangement. The series of each of these years can be had on subscribing for the year.

LAWS OF THE NEW SYDENHAM SOCIETY.

I.—The Society is instituted for the purpose of supplying certain acknowledged deficiences in the existing means of diffusing medical literature, and shall be called "THE NEW SYDENHAM SOCIETY."

II.—The Society shall carry out its objects by a succession of publications, of which the following shall be the chief:—1. Translations of Foreign Works, Papers, and Essays of merit, to be reproduced as early as practicable after their original issue. 2. British Works, Papers, Lectures, &c., which, whilst of great value, have become from any cause difficult to be obtained, excluding those of living authors. 3. Annual Volumes consisting of Reports in Abstract of the progress of the different branches of Medical and Surgical Science during the year. 4. Dictionaries of Medical Bibliography and Biography. Those included under Nos. 1 and 2 shall be held to have the first claim on the attention of the Society; and the carrying out of those under Nos. 3 and 4 shall be considered dependent upon the amount of funds which may be placed at its disposal.

III.—The Subscription constituting a Member shall be One Guinea, to be paid *in advance* on the 1st of January annually, and it shall entitle the subscriber to a copy of every work published for that year. *No books shall be issued to any Member until his subscription for the year has been paid.*

IV.—The Officers of the Society shall be elected from the Members, and shall consist of a President, sixteen Vice-Presidents, a Treasurer, a Secretary, and a Council of thirty-two, in whom the power of framing Bye-laws and of directing the affairs of the Society shall be vested. Twelve of the Council shall be provincial residents.

V.—Five Members of the Council shall form a quorum.

VI.—The Officers of the Society shall be elected by ballot at the General Anniversary Meeting of the Society. Balloting lists of Officers proposed by the Council, with blank places for such alterations as any Member may wish to make, shall be laid on the Society's table for the use of Members.

VII.—The President, Vice-Presidents, and Council, shall be eligible for re-election, except that of the Vice-Presidents four, and of the Council eight, shall retire every year.

VIII.—The Council shall appoint local Honorary Secretaries wherever they shall see fit.

IX.—The business of the President shall be to preside at the Annual and Extraordinary Meetings of the Society; in his absence one of the Vice-Presidents, or the Treasurer, or any Member of the Council chosen by the Members present, shall take the Chair.

X.—The Treasurer, or some person appointed by him, shall receive all moneys due to the Society.

XI.—The money in the hands of the Treasurer, which shall not be immediately required for the uses of the Society, shall be vested in such speedily available securities as shall be approved by the Council.

XII.—The Council shall select the Works to be published by the Society, and shall make all arrangements, pecuniary or otherwise, in regard to their publication. In the event of any Member of the Council being appointed to edit any Work for the Society, for which he is to receive pecuniary remuneration, he shall immediately cease to be a Member of the Council, and shall not be eligible for re-election till after the publication of the Work.

XIII.—The Council shall lay before the Members at each Anniversary Meeting a Report of their proceedings during the past year, and also an account of the Receipts and Expenditure of the Society; and shall further cause to be printed and circulated among the Members an abstract of such Report and Accounts immediately after such Anniversary Meeting.

XIV.—The annual Accounts of the Receipts and Expenditure of the Society shall be audited by a Committee of three Members, selected at the preceding Anniversary Meeting from among the Members at large.

XV.—The Secretary shall have the management of the general Correspondence of the Society, and of such other business as may arise in carrying out its objects.

XVI.—The local Secretaries shall further the objects of the Society in their respective districts, and shall be in communication with the metropolitan Secretary.

XVII.—The Anniversary Meeting shall be held in the same town as, and at the time of, the Annual Meeting of the British Medical Association, notice of it having been given to all Members at least a week before the day fixed on.

XVIII.—The Members generally shall be invited and encouraged to propose Works, &c., and to make any suggestions to the Council they may think likely to be useful.

XIX.—The Works of the Society shall be printed for the Members only.

XX.—No alteration in the Laws of the Society shall be made, except at a General Meeting. Notice of the alteration to be proposed must also have been laid before the Council at least a month previously.

XXI.—The Council shall have power to call a General Meeting of the Members at any time, and shall also be required to do so within three weeks, upon receiving a requisition in writing to that effect from not less than twenty Members of the Society.

XXII.—All Special General Meetings of the Society shall be held at such place as the Council may appoint.

XXIII.—The Council shall meet at least once in two months, unless by special resolution to the contrary.

GENERAL INFORMATION.

A THIRD EDITION of the VOLUMES for 1859 was printed, and a Second of that for 1860. For subsequent years the First Edition was much larger; and it is not likely that any of the Volumes will be reprinted.

Most of the stones for Plates, &c., both those for the Atlas of Skin Diseases and those for printed Volumes, have been destroyed, and will not be reproduced.

A few complete Sets of the Society's Works are on hand, and can be obtained by new Members, but the number remaining is very small.

The Society is now in its Twenty-second year, and the cost of a complete Set to the end of 1879 is Twenty-one Guineas. Arrangements have been made by which new Members can obtain single Volumes, or sets of Volumes, from the Society's stock in hand. Some of the Volumes, of which a larger surplus exists than of others, can be purchased at fixed prices (for which see list.) The Society's Agent is empowered to make special arrangements with new Members who may wish to obtain any of the past Volumes.

CARRIAGE, &c.—The Society's Works are supplied free of cost to any address in London, Edinburgh, or Dublin; but the expenses of Carriage to all other places must be borne by the Members to whom they are sent. Members wishing to receive their Volumes by Book-post can do so by prepaying the postage. Members are requested to give detailed instructions respecting the mode by which they wish their Volumes to be forwarded, and also to remember that the Society's responsibility ceases when the Book has been delivered according to the instructions given. Members wishing to receive their Works by Book-post can do so by prepaying the sum of 2*s.* 6*d.* for the year.

The Subscription is One Guinea annually, to be paid *in advance.* The best mode of sending money is by Post-office Order, payable to Mr. HENRY KING LEWIS, at the London Office; or by Cheque to the order of the Treasurer, Dr. SEDGWICK SAUNDERS. It is requested that in future all communications in reference to the payment of Subscriptions, or the issue of Books, may be made to Mr. LEWIS, the Society's Agent, and not to the Secretary.

JONATHAN HUTCHINSON,
Hon. Secretary.

15, CAVENDISH SQUARE, W.

*** Any Member wishing for additional Copies of this Report, &c., can obtain them by applying to Mr. HUTCHINSON; or to the Society's Agent, Mr. LEWIS, 136, Gower Street, W.C. The Council will be much obliged by its distribution amongst those thought likely to join the Society.

PS.—The Society's Agent is prepared to supply, at fixed prices, CASES for binding the Lexicon, and PORTFOLIOS for the reception of the Plates of Skin Diseases.

NOTICE TO NEW SUBSCRIBERS AND LOCAL SECRETARIES.

New members who subscribe for not fewer than three past years at once will be allowed to select volumes from the surplus stock to the value of one guinea without additional payment (see page 29 for list of surplus stock and prices). The like privilege will be secured each year by any Local Secretary who has not fewer than ten members on his list whose subscriptions are paid before the end of March.

LIST OF THE

OFFICERS AND MEMBERS

OF

The New Sydenham Society.

1880—81.

PRESIDENT.

Sir William W. Gull, Bart., M.D., F.R.S., D.C.L., LL.D.

VICE-PRESIDENTS.

Robert Barnes, M.D.
Sir George Burrows, F.R.S., Bart.
R. W. Falconer, M.D., D.C.L., *Bath.*
W. D. Husband, Esq., F.R.C.S., *Bournemouth.*
W. T. Gairdner, M.D., *Glasgow.*
Cæsar H. Hawkins, Esq., F.R.S.
T. Holmes, Esq.
*George Johnson, M.D., F.R.S.
Joseph Lister, Esq., F.R.S., D.C.L.
Sir James Paget, F.R.S., Bart.
*George Paget, M.D., F.R.S., *Cambridge.*
*William Rutherford, M.D., F.R.S., *Edinburgh.*
T. Grainger Stewart, M.D., *Edinburgh.*
Sir Thomas Watson, M.D., F.R.S., Bart.
Hermann Weber, M.D.

COUNCIL.

*T. Clifford Allbutt, M.D., F.R.S.

*Milner Barry, M.D., *Tunb. Wells.*

Thomas Barlow, M.D.

*Lauder Brunton, M.D., F.R.S.

W. H. Broadbent, M.D.

Thomas Buzzard, M.D.

W. Cholmeley, M.D.

William Colles, M.D., *Dublin.*

R. M. Craven, Esq., *Hull.*

J. Langdon H. Down, M.D.

Dyce Duckworth, M.D.

J. Matthews Duncan, M.D.

John Easton, M.D.

Sir Joseph Fayrer, M.D.

C. J. Hare, M.D.

G. E. Herman, M.D.

H. Macnaughten Jones, M.D., *Cork.*

J. C. Langmore, M.B.

P. W. Latham, M.D., *Cambridge.*

*Stephen Mackenzie, M.D.

J. W. Moore, M.D., *Dublin.*

*Walter Moxon, M.D.

*Wm. Roberts, M.D., *Manchester.*

Septimus W. Sibley, Esq.

J. K. Spender, M.D., *Bath.*

*Wm. Square, Esq., *Plymouth.*

*Paul Swayne, Esq., *Devonport.*

T. P. Teale, Esq., *Leeds.*

Wm. Turner, M.B., F.R.S.E., *Edinburgh.*

John Walters, M.B., *Reigate.*

James West, Esq., *Birmingham.*

TREASURER.

W. Sedgwick Saunders, M.D., 13, *Queen Street, Cheapside, E.C.*

AUDITORS.

E. Clapton, M.D. | S. Fenwick, M.D.

F. M. Corner, Esq.

HON. SECRETARY.

Jonathan Hutchinson, Esq., 15, *Cavendish Square, W.*

DEPÔT AND AGENCY.

Mr. H. K. Lewis, 136, *Gower Street, W.C.*

Those marked with an Asterisk were not in office last year.

LIST OF HON. LOCAL SECRETARIES,

And of Towns where it is desired that an appointment should be made.

The Council will be much obliged to any gentlemen willing to act as Local Secretaries in Towns where the appointment is vacant, if they will communicate with Mr. Hutchinson. *Any suggestions of suitable names will also confer a favour. The duties of Local Secretaries consist in arranging for the distribution of books, the collection of Subscriptions, and canvassing for new members.*

ENGLAND AND WALES.

Aberdare	
Abergavenny	
Aberystwith	Morris Jones, Esq.
Abingdon	Paulin Martin, Esq.
Accrington	
Acton (see Ealing)...	
Alfreton	
Alnwick	
Altrincham...	
Ampthill	
Andover	W. H. Lush, F.R.C.P. Ed. (Fyfield).
Arundel	
Ashbourne	
Ashford	
Ashton-under-Lyne	
Aylesbury	Robert Ceely, Esq.
Bacup	
Banbury	
Bangor	
Barnsley	
Barnstaple	R. Budd, Esq., M.D.
Bath	J. K. Spender, M.D.
Beaminster and Bridport	J. S. Webb, Esq.
Beaumaris, Anglesea	
Beccles	W. M. Crowfoot, Esq.
Bedford	R. H. Coombes, M.D.
Beverley	
Bethesda, Carnarvonshire ...	
Bewdley	J. Gabb, Esq.
Bideford	W. H. Ackland, M.D.

Bilston	
Birkenhead	George Walker, M.D.
Birmingham	W. Wright Wilson, Esq.
Bishop Auckland	T. A. McCullagh, Esq.
Bishop's Stortford...	
Blackburn	Matthew J. Rae, M.D.
Blackheath...	Gay Shute, Esq.
Blackpool	W. B. Ricnardson, Esq.
Bodmin	
Bolton	
Boroughbridge	
Boston, Lincolnshire	A. Mercer Adam, M.D.
Bournemouth	
Bradford, Yorkshire	T. C. Deaby, Esq.
Brecon	Talfourd Jones, M.D.
Brentwood	
Bridgend	
Bridgewater	W. L. Winterbotham, M.D.
Bridgnorth...	Alfred Mathias, Esq.
Bridlington	C. F. Hutchinson, M.D.
Brighton	Ed. Mackey, M.D.
Bristol	F. R. Cross, Esq., F.R.C.S.
Bromley, Kent	
Brompton, Kent	
Burnley, Lancashire	W. M. Coultate, Esq.
Burton-on-Trent	G. Lowe, M.D.
Bury, Lancashire	J. S. C. Yule, Esq.
Bury St. Edmunds	F. E. Image, Esq.
Buxton, Derbyshire	
Cambridge	E. Carver, M.B.
Canterbury...	James Reid, Esq.
Cardiff	
Carlisle	W, B. Page, Esq.
Carmarthen	J. Hughes, Esq.
Carnarvon	
Castleford	E. W. Kemp, Esq.
Chard	
Chatham	
Cheadle, Cheshire	
Cheadle, Staffordshire	
Chelmsford...	
Cheltenham	E. T. Wilson, M.D.
Chertsey	
Chepstow	
Chester	W. McEwen, M.D.
Chesterfield	John Carnegie, M.D.
Chichester	N. Tyacke, M.D.
Chippenham, Wilts	
Chorley, Lancashire	

Christchurch	
Cirencester...	
Colchester	E. Waylen, Esq.
Colney Hatch	
Congleton	
Coventry	J. Brown, M.D.
Cowes, Isle of Wight ...	
Crewe	J. Atkinson, Esq.
Croydon	A. Carpenter, M.D.
Darlington	J. Lawrence, M.D.
Dartford	
Dartmouth	
Deal, Kent	
Denbigh ...	
Deptford	
Derby	F. W. Wright, Esq.
Devizes	G. Waylen, Esq.
Devonport	
Dewsbury	
Diss...	T. E. Amyot, Esq.
Doncaster	J. Sykes, M.D.
Dorchester	G. Curme, Esq.
Dorking	
Douglas, Isle of Man	
Dover	Charles Parsons, M.D.
Droitwich	S. S. Roden, M.D.
Dudley	
Durham	
Ealing	J. Goodchild, Esq.
Eastbourne ...	B. Roberts, M.D.
East Grinstead ...	
East Retford ...	W. B. Pritchard, Esq.
Edmonton	
Ely	
Enfield	
Epsom	Clement Daniel, M.D.
Evesham ...	
Exeter ...	
Exmouth ...	G. W. Turnbull, M.D.
Falmouth ...	
Faversham ...	
Folkestone ...	R. L. Bowles, M.D.
Forest Hill ...	J. Bright, M.D.
Frome ...	
Gainsborough	D. Mackinder, M.D.
Gloucester ...	F. Needham, M.D.
Godalming ...	
Gosport ...	
Grantham ...	G. W. Shipman, Esq.

Gravesend	R. Innes Nisbett, Esq.
Great Grimsby	
Greenwich and Blackheath	Gay Shute, Esq.
Guernsey	B. Collenette, M.D.
Guildford	
Halifax	
Hanley	
Hanwell	
Harrow	R. N. Day, Esq.
Harrogate	G. Oliver, M.D.
Harrow-on-the-Hill	
Hartlepool	
Hastings	J. Underwood, M.D.
Haverfordwest	J. E. Brown, M.D.
Heckmondwike	F. B. Lee, Esq.
Hemel-Hempstead...	
Hereford	Thomas Turner, Esq.
Hertford	C. E. Shelley, M.D.
Hexham	
High Wycombe	
Hillingdon	
Hinkley, Leicestershire	
Hitchin	R. R. Shillitoe, Esq.
Hounslow	
Holywell	
Huddersfield	John Irving, Esq., M.B.
Hull	Kelbourne King, M.D.
Huntingdon	L. Newton, Esq.
Hyde and Marple	J. Johnson Bailey, M.D.
Ilfracombe	
Ipswich	C. W. Hammond, M.D.
Isleworth	
Jersey	
Kendal	
Kenilworth...	
Kettering	
Kidderminster	
Kingsbridge	
Kingston-upon-Thames & Surbiton	W. W. Kershaw, M.D.
Knottingley	
Lancaster	
Langport	
Launceston... ...	
Leamington ...	T. W. Thursfield, M.D.
Ledbury	
Leeds ...	F. Greenwood, Esq.
Leek	Joseph Kenny, Esq.
Leicester ...	J. Barclay, M.D.
Leominster...	

Leytonstone ...	F. W. Cooper, Esq.
Lewes	
Lichfield	H. P. Welchman, Esq.
Lincoln	T. Sympson, M.D.
Liskeard	
Liverpool	J. Muir Howie, M.B.
Llandilo	
Llandovery	D. Thomas, Esq.
Llandudno	
Llanelly	J. Raglan Thomas, Esq.
Longton, Staffordshire	
Louth	F. Fawssett, M.D.
Lowestoft	W. H. Clubbe, Esq.
Ludlow	
Luton	
Lutterworth	
Lynn	
Macclesfield	E. Woodward, Esq.
Maidenhead	
Maidstone	
Maldon, Essex	
Malton	W. T. Colby, Esq.
Malvern	W. C. West, M.D.
Manchester	J. Chadwick Peatson, M.D.
Mansfield	
Market Drayton	
Marlborough, Wilts	
Martock	
Merthyr Tydvil	
Middlesboro'-on-Tees	J. Hedley, Esq.
Mold	
Monmouth	
Moreton-in-Marsh	
Morpeth	
Newark-upon-Trent	F. H. Appleby, Esq.
Newbury, Berks	J. Bunny, M.D.
Newcastle-under-Lyne	
Newcastle upon Tyne	Thomas Oliver, M.D.
New Malton, Yorkshire	
Newmarket, Cambridgeshire ...	
Newport, Hants	
Newport, Mon.	W. W. Morgan, M.D.
Newton Abbot	
Newton-le-Willows	Frederick Noon, Esq.
Northampton	Charles Jewel Evans, Esq.
North Shields	Robert Peart, M.D.
Northwich	
Norwich	Hayness Robinson, Esq.
Nottingham ...	W. H. Ransom, M.D.

Odiham	J. M'Intyre, M.D.
Oldham	T. Platt, Esq.
Oswestry	
Otley	
Oxford	A. Winkfield, Esq.
Penrith	
Penzance	J. B. Montgomery, M.D.
Peterborough	
Petersfield	
Plymouth	Connell Whipple, Esq.
Pontefract	
Poole	
Pontypool	S. B. Mason, Esq.
Portsmouth (*see* Southsea) ...	
Preston	R. Allen, Esq.
Putney	
Ramsgate and I. of Thanet ...	E. Walford, Esq.
Reading	
Redruth	
Reigate	J. Walters, M.D.
Richmond, Surrey	E. Fenn, M.D.
Ripon	
Rochdale	R. M. Pooley, Esq.
Rochester (& Chatham & Strood)...	
Rochford	T. King, M.D.
Romford, Essex	
Ross	
Rotherham...	H. D. Foote, M.D.
Rugby	
Rugeley	
Ryde, I. of Wight...	
Rye, Sussex	
Saffron Walden	H. Stear, Esq.
St. Albans	
St. Austell...	
St. Helen's, Lancashire	E. P. Twyford, M.D.
St. Ives	W. R. Grove, M.D.
Salisbury	W. D. Wilkes, Esq.
Scarborough	R. B. Cooke, Esq.
Shaftesbury	
Sheerness	E. Swales, Esq.
Sheffield	M. Martin de Bartolomé, M.D.
Shepton Mallet	
Sherborne ...	
Shirley, Hants	
Shrewsbury	E. Andrew, M.D.
Sidmouth ...	
Skipton ...	
Sleaford ...	

Smethwick	
Snaith	
Southampton	R. W. Griffin, M.D.
Southmolton	
Southport	
Southsea	W. H. Axford, M.D.
South Shields	J. Frain, M.D.
Spalding	E. Morris, M.D.
Stafford	S. Cookson, M.D.
Stalybridge	
Stamford	W. Newman, M.D.
Stockport	J. A. Ball, Esq.
Stockton-on-Tees	W. H. Oliver, Esq.
Stoke-on-Trent (Potteries) ...	Samuel Johnson, M.D.
Stourbridge	A. Freer, Esq.
Stratford, Essex	
Stratford-on-Avon	J. J. Nason, M.B.
Stroud, Gloucestershire	
Sunderland	M. Douglas, Esq.
Surbiton	W. W. Kershaw, M.D.
Swansea	T. D. Griffiths, M.B.
Swindon	G. M. Swinhoe, Esq.
Sydenham (*see* Forest Hill) ...	
Taunton	W. Liddon, M.B.
Tavistock	
Teignmouth	
Tenby	
Tewkesbury	
Thetford	P. Minns, M.D.
Thirsk	W. H. Ryott, M.D.
Tiverton	
Torquay	
Totnes	
Tottenham	E. H. May, M.D.
Truro	C. Sharp, Esq.
Tunbridge	
Tunbridge Wells	J. Milner Barry, M.D.
Twickenham	
Ulverston	
Uxbridge	G. H. Macnamara, Esq.
Ventnor	
Wakefield	F. H. Wood, Esq.
Wallingford	C. A. Barrett, Esq.
Walsall	
Warminster	
Warrington	J. H. Gornall, Esq.
Warwick	
Wednesbury	
Watford	

Wellington, Somerset ... ...
Wellington, Salop... ... ...
Wells ... ... J. G. French, Esq.
Welshpool, Montgomeryshire ...
Wem ... ...
West Bromwich ... ... J. Manley, Esq.
Weston-super-Mare ... ... R. Alford, Esq.
Weymouth ... ...
Whitby ... ... John Yeoman, M.D.
Whitehaven ... ... J. F. I'Anson, M.D.
Wigan ... ... G. G. Tatham, M.D.
Wimbledon ... ...
Wimborne ... ...
Winchester ... ... F. J. Butler, M.D.
Windsor ... ... J. Ellison, M.D.
Wirksworth ... ...
Wisbeach ... ...
Witney ... ... A. Batt, M.D.
Wolverhampton ... ... Vincent Jackson, Esq.
Woodbridge ... ...
Woodford ... ...
Woolwich ... ... R. Mason, Esq.
Worcester ... ...
Worksop ... ...
Worthing ... ... W. J. Harris, Esq.
Wrexham ... ... E. Davies, M.D.
Yarmouth ... ... C. Palmer, Esq.
Yeovil ... ...
York ... ... G. Shann, Esq.

SCOTLAND.

Aberdeen ... ... John Wight, M.D.
Ayr... ... ... G. McKerrow, M.D.
Banff ... ... J. Barclay, M.D.
Brechin ... ...
Coldstream... ... ...
Cupar, Fife ... ... J. W. Reid Mackie, M.D.
Dumfries ... ...
Dundee and Forfar ...
Dumfermline ... ...
Edinburgh... ... ... W. Husband, M.D.
Elgin ... ... G. Duff, M.D.
Glasgow ... ... J. W. Anderson, M.D.
Greenock ... ... James Wallace, M.D.
Haddington ... ... T. Howden, junr., M.D.
Hamilton ... ...
Helensburg .. ..
Inverness ... ...
Kilmarnock ... ... — Macfarlane, M.D.

Leith	James Struthers, M.D.
Lerwick (Shetland)	
Linlithgow	G. Hunter, M.D.
Lochgilphead	J. Rutherford, M.D.
Montrose	James C. Howden, M.D.
Paisley	D. Taylor, M.D.
Peebles	
Perth	D. H. Stirling, M.D.
Rothesay	
St. Andrews, Fife	
Stirling	Charles Gibson, M.D.
Thurso	
Wishawton	

IRELAND.

Ardee	Thomas J. Moore, M.D.
Armagh	Thomas Cuming, M.D.
Ballinasloe	
Belfast	(Books per Mr. Greer.)
Carlow	
Carrick-on-Suir	J. Martin, M.D.
Cashel	
Cavan	W. Malcolmson, M.D.
Clonmel	
Cork	E. Finn, M.D.
Dublin	J, W. Moore, M.D.
Dundalk	H. Macdonnell, M.D.
Ennis	
Enniskillen	
Galway	
Killarney	
Kilkenny	
Kingstown	W. O'Brein Adams, M.D.
Lifford	R. Little, M.B.
Limerick	T. Kane, M.B.
Letterkenny	
Lisburn, Antrim	
Listowel	
Londonderry	W. Bernard, M.D.
Mallow	W. J. Galway, M.D.
Moate	
Monaghan	
Mullingar	
Nenagh	
New Ross	
Newry	
Parsonstown	
Queenstown	
Roscommon	J. Harrison, M.D.

Rosstrevor	T. A. Vesey, M.B.
Rothmines	
Roscrea	
Sligo	
Thurles	
Tralee	
Tullamore	
Waterford	
Westport	
Wexford	H. H. Boxwell, M.D.
Youghal	

PARIS.	*FLORENCE.*
S. Pozzi, M.D.	David Young, M.D.

INDIA.

Calcutta	
Madras	E. F. Brockman, M.D.
Bombay	
Lahore	
Moulton	Joseph Blood, M.B.

S. AUSTRALIA.

Adelaide ... H. Whittell, M.D.

VICTORIA.

Melbourne ... ... Edward Barker, M.D.

NEW SOUTH WALES.

Sydney ... James Spark, M.D.

NEW ZEALAND.

Christchurch ...	J. Irving, M.D.
Nelson	
Napier, Hawkes' Bay	F. E. De Lisle, M.D.

QUEENSLAND.

Toowoomba, Brisbane ... S. Flood, M.D.

CANADA.

Montreal ...

UNITED STATES.

Abingdon, Ill.	Madison Reece, M.D.
Cincinnati	Messrs. R. Clarke & Co.
New York	Messrs. Wood & Co.
Boston	R. H. Salter, M.D.
Philadelphia	Richard J. Dunglison, M.D. (Mr. Presley Blakiston).
Baltimore	

BARBADOES.

Robert R. Walcott, M.D.

DEMERARA.

JAPAN.

Yokohama and Yeddo S. Eldridge, M.D.

General Secretary (Hon.)

JONATHAN HUTCHINSON, Esq., 15, Cavendish Square, London, W.

Agency and Depot for Books.

Mr. H. K. LEWIS, 136, Gower Street, London.

MEMBERS.

Aberdare	Jones, Evan, *Tymawr*
Abergavenny	Steel, S. H.
	Glendinning, J., M.D.
Aberystwith	Jones, Morris, *Loc. Sec.*
Abingdon	Martin, Paulin, *Loc. Sec.*
Addiscombe	Smith, S. Parsons, M.D.
Aldershot	Aldershot Medical Book Club
Alford	Handsley, T. A.
Alconbury Hill *see* Huntingdon	
Alnwick	Rowlands, W., M.B.
Alresford *see* Winchester	
Alton *see* Odiham	
Alverstoke	Tracy, S. J.
Andover	Lush, W., M.D., *Loc. Sec.*
Anerley	Partridge, S. B.
	Turner, J. Sydney
Arundel	
Ashburton	Fraser, W. J., B.M.
Ashcott	Bridgewater
Ashford	
Aylesbury	Ceely, Robert, *Loc. Sec.*
	Hooper, Charles, *Aylesbury.*
	Dickson, J. D., M.D.
	Wilcox, R. W., *Aylesbury.*
	Bond, W. J., *The Grove Brill*
Bacup	Stewart, W., M.D.
Bamburgh	Burman, C. Clark, M.D.
Bampton	Atkinson, J. P., M.D.
Bakewell	Fentem, P. S., M.D.
Banbury	
Balham	Burn, W. B.
Barnet	Harnett, W. J., M.D.
	Perigal, A., M.D.
Barnard Castle...........	Mitchell, J., M.D.
	Atkinson, J.
Barnsley....................	Jackson, A.
Barnstaple.................	Budd, R., M.D.
	Harper, J.

Barrow-in-Furness	Stark, P. W., M.D.
Basingstoke	Andrews, S.
	Miller, F., L.R.C.P.
Bath	Spender, J. K., M.D., *Loc. Sec.*
	Budd, S. P.
	Falconer, W. R., M.D.
	Fowler, R.
	Fox, A. W., M.B.
	Hensley, H., M.D.
	Lawrence, G. E., *Lincombe*
	Mason, F.
	Michael, D.
	Wyndowe, S. J., M.D.
Beccles	Crowfoot, W. M., *Loc. Sec.*
	Metcalfe, B. J., M.D.
Beaminister *see* Bridport	
Beckenham	Porter, R. V.
	Stillwell, S., M.D.
Bedford	Coombs, R. H., M.D., *Loc. Sec.*
	Goldsmith, G. P., M.D.
Bedlington	
Bellingham..................	
Belper	Gaylor, E., L.R.C.P.
Bethesda......................	
Beverley	
Bewdley	Gabb, John, *Loc. Sec.*
	Greensill, J. N., *Great Witley*
Bideford......................	Ackland, W. H., M.D., *Loc. Sec.*
Billericay	Carter, Fredk.
Bexley Heath	Barrington, N. W., M.D.
Birkenhead	Walker, G., M.D., *Loc. Sec.*
	Braidwood, P. M., M.D.
	Byerley, Isaac, *Seacombe*
	Forbes, D., M.D., *Rockferry*
	Harris, A. C. E., M.D.
	Lambert, J., M.D.
	Laidlaw, W. C., M.D., *Tranmere*
	Main, W., M.D., *New Ferry*
	Robson, John, M.D., *Rockferry*
	Spratley, S., *Rockferry*
Birmingham.............. ..	Wilson, W. Wright, F.R.C.S.E., *L. Sec.*
	Archer, John, *Edgbaston*
	Baker, A.
	Bassett, J.
	Berry, S.
	Borough of Birmingham Central Free Library
	Clay, J.
	Drummond, A., M.B.

Birmingham, *continued* ...	Evans, T. D. F.
	Evans, G. H., M.D.
	Foster, B.
	Goodall, W. P.
	Hadley, Clement
	Jones, G.
	Keyworth, J. W., M.D.
	Medical Book Society: G. Jones, *Sec.*
	Russell, J., M.D.
	Savage, T., M.D.
	Sawyer, J., M.D.
	West, J. J.
	Lloyd, G. J.
Bishop Auckland	
Blackburn	Rae, M. I., M.D., *Loc. Sec.*
	Davidson, I. R., M.D.
Blackheath	Shute, Gay, *Loc. Sec.*
	Burton, J. M.
	Forsyth, Alex., M.D., *Greenwich*
	Noyes, H. G., M.D., *Lee*
	Peacey, W., *Lewisham*
	Pope, A. C., M.D., *Lee Park*
	Purvis, P., M.D.
	Roper, A.
	Steel, C., *Lewisham*
	Sturton, H. W. S., *Greenwich*
	Williams, —., M.D., *Burnt Ash*
	Steel, R., *Greenwich*
	Burton, J. S.
Blackpool	Richardson, W. B., *Loc. Sec.*
	Hardman, W., M.D.
Blandford *see* Dorchester	
Blankland	Montgomery, W.
Bletchingley	
Bloxham	
Bolton........................	Bolton Medical Society: C. Macfie, M.D.
	Livy, J., M.D.
Bootle *see* Liverpool	
Boro'bridge	Sedgwick, J., M.D.
Boston........................	Adam A. Mercer, M.D.
	Lamphier, R., M.D., *Alford*
	Pilcher, W. J.
	Walker, J. W., M.D., *Spilsby*
Bournemouth	Drury, W. V., M.D.
	Husband, W. D., M.D.
	Bournemouth Medical Society: per W. Salt, M.D.
	Woodroffe, J. F., M.D.

Bowdon …………………	
Box …………………	
Bradford-on-Avon ………	Adye, W., M.D.
Bradford…………………	Denby, T. C., *Loc. Sec.*
	Appleyard, J.
	Brown, S., M.D.
	Foster, J.
	Lodge, S.
	Miall, Philip
	Munro, A. B., M.D.
	Infirmary Library
	Leeson, J. F., M.D., *Ilkley*
	Spenser, H.
Braintree ………………	Harrison, J.
Brampton………………	Wotherspoon, T. A., M.D.
Brecon…………………	Jones, Talfourd, M.D., *Loc. Sec.*
	North, John
	Williams, J.
Brentford ………………	
Brentwood ………………	Amsden, G., M.D.
Bridgend………………	M'Cracken, I., M.B.
Bridgnorth ………………	Mathias, Alfred, *Loc. Sec.*
	Colles, A.
	Thursfield, W., jun.
Bridgwater………………	Winterbotham, W. L., M.B., *Loc. Sec.*
	Cornwall, J., *Ashcott*
	Parsons, J.
	Todd, W. J., *North Petherton*
Bridlington	Hutchinson, C. F., M.D., *Loc. Soc.*
	Brett, F. C.
	Savile, R., M.D., *Nafferton*
Bridport ………………	Webb, J. S., *Loc. Sec.*, *Beaminster*
	Daniell, W. J., *Beaminster*
	Denzelow, Defros, M.D.
	Hay, W. H., M.D.
Brighton	Mackey, Edward, M.D., *Loc. Sec.*
	Furner, E. F.
	Humphrey, F. A.
	Library of Medical Society: per Rogers, R. J.
	Smith, Heckstall
	Vipan, C.
Brigg ………………	
Bristol ………………	Cross, F. R., F.R.C.S., *Loc. Sec.*, *Clifton*
	Beddoe, J., M.D., *Clifton*
	Belfield, C. W., L.R.C.P.
	Brittain, F., M.D., *Clifton*
	Bristol Medical Library
	Bristol General Hospital Library

BRISTOL, *continued*	Clark, Mitchell, *Clifton*
	Coe, R. W.
	Crossman, E., *Hambrook*
	Daubeney, —., M.D.
	Daviss, David
	Fendick, R.
	Fox, E. Long, M.D., *Clifton*
	Fox, C. H., M.D., *Brislington*
	Goodeve, E., M.D., *Stoke Bishop*
	Hawkins, Cæsar F.
	Lansdowne, J. G., *Clifton*
	Marshall, H., M.D., *Clifton*
	Newstead, J., *Clifton*
	Phelps, W. H.
	Pritchard, Augustus, *Clifton*
	Ring, C. Gore, L.R.C.P.E., *Clifton*
	Royal Infirmary Library
	Skerritt, E. M., M.D.
	Smith, R. Shingleton
	Spencer, W. H., *Clifton*
	Chillcott, J. E.
BROADMOOR	Orange, W., M.D.
BROMLEY	Ilott, J. W.
BROMYARD	Powell, W.
BROONSGROVE	Wood, R.
BROUGHTON	Clapperton, James
BROOK GREEN	
BUCKFASTLEIGH	Ubsdell, H.
BUCKINGHAM..................	
BUDE	Lyle, Thomas, M.D.
BUILTH	
BURNLEY	COULTATE, W. M., *Loc. Sec.*
	Briggs, H., M.D.
	Anningson, J. W.
	Herron, J., M.D.
	O'Sullivan, D.
Burntwood *see* Lichfield	
BURTON-ON-TRENT	LOWE, G., M.D., *Loc. Sec.*
	Mason, P. B.
	Wolfenden, I. W., *Tutbury*
BURY	YULE, J. S. C., *Loc. Sec.*
	Crompton, F.
	Fletcher, A., M.D.
	Mellor, J.
	Nuttall, W.
	Galloway, J., *Ramsbotham*
BURY ST. EDMUMDS.........	IMAGE, F. E., *Loc. Sec.*
Burwell *see* Ely	

Buxton	
Cambridge	Carver, E. M., *Loc. Sec.*
	Balding, M., *Royston*
	Collier, —.
	Drosier, W. H., M.D.
	Hills, F. Hyde
	Latham, P. W., M.D.
	Paget, J. E., M.D.
	Perkins, J., LL.D.
	Pyne, R., *Royston*
	University Library: per H. Bradshaw
Cannock	Taylor, Moses
	Mockett, G. T.
Canterbury	Reid, Jas., *Loc. Sec.*
	East Kent Medical Library
	Rigden, Brian
	Shaw, W.
	Sadler, H. G.
	Wacher, F., *Kingsbridge*
Capel	Jardine, J. Lee
Cardiff	Cardiff Medical Society: Horder, T. G.
	Evans, Maurice G.
Carlisle	Page, W. B., *Loc. Sec.*
	Arras, W., M.D.
	Campbell, J. A., M.D.
	Hair, P., M.D.
	Lockie, Stewart, M.D.
	M'Bean, R. B., M.D., *Annan*
	Moffat, P., M.D., *Dalston*
	Pearson, J., M.D., *Maryport*
Carmarthen	Hughes, J., F.R.C.S.
	Rowlands, J., F.R.C.S.
Carnarvon	
Castleford	Kemp, E. W., *Loc. Sec.*
Caterham Valley	Barrow, Gerald, E.
Catterick	Cockcroft, William
Chapel-Allerton *see* Leeds	
Chard	
Chapeltown *see* Sheffield	
Chatham *see* Rochester	
Chelmsford	
	Wheeler, Daniel
	Bodkin, W.
Cheltenham	Wilson, E. T., M.D., *Loc. Sec.*
	Askwith, R., M.D.
	Bubb, T.
	Carden, A.
	Kerr, C. B., M.D.
	Kilgour, J. S., M.D.

CHELTENHAM, *continued* ...	Ferguson, G. B.
	Howard, H. H.
	Bevan, J. P., L.R.C.P.
	Bennett, C. J.
	Medical Society
	Thorpe, D., M.D.
	Walters, C. A.
	Winterbotham, L.
	Smith, T., M.D.
CHEPSTOW	
CHERTSEY	Shurlock, Mainwaring
CHESTER	M'EWEN, W., M.D., *Loc. Sec.*
	Davis-Colley, T., M.D.
	Dobie, W., M.D.
	Haining, W., M.D.
	Hamilton, Alex.
	Infirmary Library
	Jackson, R. A., M.D., *Little Sutton*
	Moreton, J. E., *Tarvin*
	Roberts, J., M.D.
	Russell, D., M.D., *Neston*
	Taylor, Jas.
	Waters, E., M.D.
	Weaver, F. P., M.D., *Frodsham*
	Williams, W., M.D., *Mold*
	Woodward, J. E.
CHESTERFIELD	CARNEGIE, J., M.D., *Loc. Sec.*
	Booth, C., M.D.
	Foulds, S.
	Jeffreys, R.
	Manson, D., M.D.
	Robinson, H.
	Rose, J., M.D.
Chester-le-Street *see* Newcastle-on-Tyne	
CHICHESTER	TYACHE, N., M.D., *Loc. Sec.*
	Buckell, L., M.D.
	Freeland, F.
CHIPPENHAM	Crisp, J. H., *Lacock*
CHIPPING NORTON	Hopgood, Thomas
CHISWICK	Leigh, W., M.D.
CHUDLEIGH	Lillies, G. W., M.D.
	Watson, J. Adams, L.R.C.P.E.
CINDERFORD	Whatmough, Charles, M.D.
CIRENCESTER	Wilson, C. W., M.D.
CLEVEDON.....................	Davis, T., M.D.
	Skinner, S., M.B.
CLEOBURY MORTIMER	
Clifton *see* Bristol	
COCKERMOUTH	Dodgson, H., M.D.

COGGLESHALL	Giles, Harold
COLCHESTER	WAYLEN, —., M.B., *Loc. Sec.*
	Colchester Medical Society
	Bree, C. R., M.D.
	Salter, J. H., *Tolleshunt D'Arcy*
Collingham *see* Newark-on-Trent	
COLNEY HATCH	Marshall, W. G.
CONGLETON	Beales, R., M.D.
CORBRIDGE	M'Lean, H., M.D.
CORSHAM	
Cotham *see* Bristol	
COSHAM	Martin, H. A., M.D.
COVENTRY	BROWN, J., M.D., *Loc. Sec.*
	Millerchip, T., M.D.
	Partridge, —, M.D., *Meriden*
COWES	Hoffmeister, W. C., M.D.
	Jones, John
COWFOLD	Graveley, T.
CRANBROOK	Joyce, T., M.D.
CREWE	Atkinson, J.
	Vaughan, W. E.
CROMER	M'Kelvie, R., M.D.
CRONDALL......................	Burroughs, T. W.
CROYDON	CARPENTER, A., M.D., *Loc. Sec.*
	Cleaver, H. A.
	Lanchester, H., M.D.
	Richardson, J. A.
	Stevens, A. F., *Wallington*
	Thompson, H. G., M.D.
	Tompkins, C. P., *Beddington*
CWM-AVON	James, J. R.
DARLINGTON	LAWRENCE, J., M.D., *Loc. Sec.*
	Cockcroft, G.
	Fothergill, J. R.
	Mackie, J., *Heighington*
	Atkinson, J., M.D., *Barnard Castle*
	Arrowsmith, W. H., M.D.
DALTON-IN-FURNESS.........	Balbirnie, J. P., L.R.C.P.
DARTFORD......................	Weddell, J. C., M.D.
	M'Naul, H. H., M.B.
DARTMOUTH	RICHARDSON, RALPH, M.D. (2 copies), *Loc. Sec.*
	Cockin, J., R.N.
DAVENTRY	
DAWLEY GREEN	Soame, C. B. H.
DAWLISH	Baker, A. DeWinter
DEAL	Davey, R. S., M,D.
	Hughes, D.

Denbigh	Roberts, G. M., M.D.
Deptford	Cope, Ricardo
Derby	Wright, F. W., *Loc. Sec.*
	Baker, J. W.
	Borough, F.
	Cooper, C. A., M.D., *Ilkeston*
	Gentles, T. J.
	Iliffe, Frank
	Ogle, W., M.D.
	Walker, B., *Spondon*
	Wigg, T. Carter
Devizes	Waylen, G., *Loc. Sec.*
	Anstie, T. B.
	Carless, E. N., M.D.
	Carter, C. H., M.D., *Pewsey*
	Hitchcock, C., M.D., *Market Lavington*
	Langdon, H. W., M.R.C.S.
Devonport	Medical Society: per Mr. Swiss
Diss	Amyot, T. E.
Doncaster	Sykes, J., M.D., *Loc. Sec.*
	Arbuckle, H. W., M.D., *Thorne*
	Kenyon, J., *Hooton Paynell*
	Phillips, G. G., *Tick Hill*
	Stone, E., M.D., *Knottingley*
	Storrs, Robert
Dorchester	Curme, G., *Loc. Sec.*
	Bacot, W. G., *Blandford*
	County Hospital Medical Library
	Clapcott, J., *Evershott*
	Emson, Alfred
	Evans, G. M., *Bridport*
	Good, J.
Dorking	
Douglas (Isle of Man)	
Dover	Parsons, Chas., M.D., *Loc. Sec.*
	Astley, E. F.
	Walter, Clement
Dowlais	Jones, Lewis, *Burryport*
Downham Market *see* Ely	
Droitwich	Roden, S. S., M.D., *Loc. Sec.*
	Whitely, Ed. A.
Dudley	Bodington, G. F., M.D., *Kingswinford*
Durham	Jepson, E. C.
Ealing	Goodchild, John, *Loc. Sec.*
	Brown, G. D.
	Christie, T. B., M.D.
	Lingham, H. B., *Acton*
Earl's Shilton *see* Leicester	

EASTBOURNE.................	ROBERTS, BRANSBY, M.D., *Loc. Sec.*
	Mundee, G.
	Gould, E. Gardiner, L.K.Q.C.P.I.
EAST DEREHAM	Vincent, J., M.D.
EAST MOLESEY	Skimming, R., M.D.
EAST RETFORD	PRITCHARD, W. B., *Loc. Sec.*
Eccles *see* Manchester	
Eckington *see* Sheffield	
Egremont *see* Whitehaven	
ELLESMERE	Roe, J. W., M.D.
ELY	Lucas, Thomas, *Burwell*
	Wales, T. G., *Downham Market*
EMSWORTH	Orsborne, John, M.D.
ENFIELD	Mugliston, G., M.D.
	Ridge, J. J., M.D.
Epping *see* Harlow	
EPSOM	DANIELL, W. C., M.D., *Loc. Sec.*
	Jones, A. O'Brien
EVESHAM	
Evershot *see* Dorchester	
EWELL........................	Barnes, G. R., M.D.
EXETER	Devon and Exeter Hospital: J. Bankhart
	Shapter, Lewis, M.D.
	Saunders, G., M.D., *Exminster*
	Rogers, N., M.D.
Exminster *see* Exeter	
EXMOUTH.....................	TURNBULL, G. W., M.D., *Loc. Sec.*
	Bryce, J. B.
	Ewen, A. B.
	Walker, R., M.D., *Budleigh Salterton*
EYE	Barnes, Edgar, M.D.
FAKENHAM	Chambers, T. S., L.R.C.P.E.
FALMOUTH	Cann, Thomas
FARNINGHAM................	Ashurst, W. R., M.D.
FAVERSHAM	Powell, Scudamore A., M.B.
FERRYHILL	Clark, H., M.D.
FESTINIOG	
FINCHLEY.....................	Hochee, J.
FISHGUARD	Thomas, J. B., M.D.
FOLKESTONE	BOWLES, R. L., M.D., *Loc. Sec.*
	Bateman, W.
	Eastes, Sylvester
	Fitzgerald, C. E., M.D.
	Hackney, J., *Hythe*
	Tyson, W. T.
FOREST HILL	BRIGHT, J., M.D., *Loc. Sec.*
	Collingridge, W.

Frome .... Cockey, Edmund
Fyfield

Gainsborough .... Mackinder, D., M.D., *Loc. Sec.*
Gateshead *see* Newcastle-on Tyne
Gloucester.................. Needham, F., M.D., *Loc. Sec.*
Waddy, H. E.
Godalming *see* Guildford
Gomersal *see* Leeds
Goole .... Bramwell, J. M.
Gosforth *see* Whitehaven
Gosport .... Mumby, R. H.
Mumby, L. P.
Grantham .... Shipman, G. W., *Loc. Sec.*
Ashby, A., M.D.
Paterson, —, M.D.
Gravesend .... Nisbett, R. J., *Loc. Sec.*
Armstrong, J., M.D.
Birdwood, R. A.
Pinching, C. J.
Great Witley *see* Bewdley
Great Haddon .... Willans, W. Blundell
Great Wakering .... Raper, W. A., M.D.
Greenfield *see* Llanelly
Greenwich *see* Blackheath
Grimsby .... Stephenson, G. S., M.B.
Guernsey .... Collinette, B., M.D., *Loc. Sec.*
Corbin, M. A. B., *St. Peters*
Guildford .... Lafargue, G. F. H., M.D., *Godalming*
Morton, J., M.D.
Surrey County Hospital
Yate, T., *Godalming*
Sutcliff, J. H., *Ripley*

Hales Owen .... Phillips, D. W.
Halifax .... Waite, W.
Halstead.................... Kellett, R. G., M.D.
Hambrook *see* Bristol
Hampton *see* Kingston-on-Thames
Hanwell..................... Richards, J. P.
Harewood *see* Leeds
Harlow .... Day, R. N., *Loc. Sec.*
Clegg, J., *Epping*
Harpurhey *see* Manchester
Harrowgate .... Oliver, G., M.D., *Loc. Sec.*
Hartley, —, M.D., *Thirsk*
Hartley, J. F.
Hunt, H. J.
Johnson, C. J. B., M.D., *Wetherby*

Harrow-on-the-Hill ...... Bridgwater, T., M.D.
Kiernan, L., M.D.
Hartlepool (West) ...... Mackeehnie, D., M.D.
Haslingden .... Aspinall, W., M.D.
Hastings .... Underwood, J., M.D., *Loc. Sec.*
Adey, Charles, M.D., *St. Leonards*
East Sussex Medico-Chirurgical Society
Heath, R., M.D.
Humphreys, J. H.
Trollope, T., M.D.
Havant
Haverfordwest .... Brown, J. D., M.D.
Hawkshead .... Parsons, G., M.D.
Hay .... Applebe, E. A.
Hincks, T. S. H., M.D.
Hayward's Heath ....
Heckmondwike
Heighington *see* Darlington
Hemel-Hempstead .... Steele, Russell
Hemsworth .... Leek, Thomas, M.D.
Henfield...................... Lewis, C. F., L.R.C.P.
Henley-on-Thames......... Jeston, R. P.
Hereford .... Turner, Thomas, *Loc. Sec.*
Bull, H. G., M.D.
Lingen, Charles, M.D.
Mason, Thomas, *Wellington*
Price, J.: for Book Society
Thomason, R.
Hertford .... Shelley, C. E., M.D.
Hexham *see* Newcastle-on-Tyne
Heywood .... Jameson, G. H., M.D.
High Barnet *see* Barnet
High Wycombe .... Turner, John
Higham Ferrars *see* Northampton
Hitchin .... Shillitoe, R. R., *Loc. Sec.*
Hogsthorpe .... Rainey, W. B., M.D.
Holbeach *see* Spalding
Hollingwood *see* Manchester
Hollingworth .... Pomfret, H. L.
Wyld, Harold, M.D.
Hooten Pagnell *see* Doncaster
Holt .... Skrimshire, J. T.
Hales, R. T., M.D.
Honley .... Smailes, T., L.R.C.P.
Hoxne .... Phillips, J. Dudley
Huddersfield .... Irving, J., M.D., *Loc. Sec.*
Cameron, J. S., M.D.
Haig, Thomas, *Meltham*
Hull .... King, R., M.D., *Loc. Sec.*

HULL, *continued*	Craven, R. M.
	Daly, O., M.D.
	Dosser, J.
	Gibson, J. H.
	Hardey, E. P.
	Hewson, W. K.
	Infirmary Library
	Locking, J. A.
	Merson, J., M.D.
	Nicholson, R. H. B.
	Plaxton, W., M.D.
	Rockcliffe, W. C., M.D.
	Sharpe, R.
HUNTINGDON..................	NEWTON, L., *Loc. Sec.*
	Book Society
	Calcott, Lewis B., *Oundle*
HYTHE	
Ibstock *see* Leicester	
Ilkley *see* Otley	
IPSWICH	HAMMOND, C. W., M.D., *Loc. Sec.*
	Book Society
	Elliston, G. S., M.D.
	Benham, H. S., M.D.
	Casley, —, M.D.
	Sampson, G. G.
JERSEY........................	Godfray, A., M.D.
KEIGHLEY.....................	Chaffers, E.
	Roberts, A.
	Jack, W., M.D.
KETTERING	Dryland, J. W.
KIDDERMINSTER	
Kingsbridge *see* Plymouth	
Kingston *see* Surbiton	
KINTBURY.....................	Lidderdale, J.
KNARESBOROUGH	
LANCASTER	Shuttleworth, G. E., M.D.
	Cassidy, D. M'K., M.D.
LANGPORT	
LANGLEY MOOR	Stewart, R., M.D.
LEAMINGTON	THURSFIELD, T. W., M.D., *Loc. Sec.*
	Baker, R. L., M.D.
	Clark, J. F.
	Haynes, —, M.D.
	Morris, J.
	Shapley, H. T., M.B,

LEAMINGTON, *continued* ...	Thorne, F.
LEEDS	GREENWOOD, F., *Loc. Sec.*
	Beverley, W. H.
	Carter, J. B., *Chapel Allerton*
	Clare, W.
	Gisburn, I. I. W.
	Greenwood, J. W., *Ossett*
	Hallilay, J.
	Jessop, T. R.
	Nevitt, J. G., *Chapel Allerton*
	Oglesby, R. P.
	Price, W. N.
	Ramsbotham, S. H., M.D.
	Ramskill, J.
	Scattergood, T.
	School of Medicine
	Taylor, G. S.
	Teale, T. P., M.A., M.B.
	Wheelhouse, C. G.
LEEK	KENNY, JOSEPH, *Loc. Sec.*
	Dakeyne, T. E., M.D.
LEICESTER	BARCLAY, J., M.D., *Loc. Sec.*
	Blunt, T., M.D.
	Cooper, C. W., M.B.
	Eddowes, J. H., M.D., *Loughboro'*
	Fulshaw, R., *Earl's Shilton*
	Grant, F., *Market Harboro'*
	Hatchett, J., M.D., *Ravenstone*
	Johnson, T. S., *Great Peatling*
	Leicester Infirmary Library
	Leicester Medical Book Society
	Pearce, G. A.
	Thomas, R. H., *Ibstock*
	Thompson, C., M.D.
	Waterhouse, J. B., *Great Peatling*
	Wood, J. A., *Sheepshed*
	Wright, J., *Markfield*
LEIGHTON BUZZARD	
LEIGH	Hall, John
LEATHERHEAD	Hurd-Wood, J., M.D.
LEOMINSTER	
LEWISHAM	Roberts, H. W.
LEYTONSTONE	Cooper, F. W.
Leyland *see* Preston	
LICHFIELD	WELCHMAN, H. P., *Loc. Sec.*
	Davis, R. A., M.D., *Burntwood*
	Morgan, Herbert
	Guthrie, G., M.D., *Burntwood*
LINCOLN	SYMPSON, T., M.D., *Loc. Sec.*

Lincoln, *continued*	Harrison, C., M.D.
	Lowe, G. M., M.B.
	O'Neil, W., M.D.
	Wilkinson, T. H.
Linslade	Storey, W., M.D.
Liverpool	Howie, J. Muir, M.B., *Loc. Sec.*
	Bailey, F. J.
	Barr, J., M.D., *Everton*
	Batty, W. E. L.
	Bickersteth, E. R.
	Bligh, —, M.D.
	Campbell, Macfie
	Cregeen, J. N.
	Dawson, T.
	Drysdale, J. J., M.D.
	Fraser, D. M.
	Gee, R., M.D.
	Gill, G.
	Glynn, M.D.
	Greenwood, H.
	Grimsdale, T. F.
	Harvey, —, M.D., *Wavertree*
	Horton, A., M.D.
	Hicks, J. S.
	Hutchinson, J. B.
	Limrick, W. A., M.D.
	Macbeth, J., M.D.
	Manifold, W. H.
	Medical Institution
	Paul, F. T.
	Parker, E.
	Rogers, T. L., M.D.
	School of Medicine
	Smith, J. K., M.D.
	Skinner, T.
	Taylor, J. S., M.D.
	Warburton, J. W., M.D.
	Walker, G. E.
	Walker, G., M.D., *Bootle*
	Watson, T. B., M.D., *New Brighton*
	Whittle, E., M.D.
	Wigglesworth, A., *Everton*
Llandudno	Nicol, J., M.D.
Llandovery.................	Thomas, D., *Loc. Sec.*
Llanrwst	Jones, T. E.
Llanelly....................	Thomas, J. Raglan, *Loc. Sec.*
	Jones, T. M., *Loughor*
	Samuel, Richard, *Greenfields*

LONDON.

Adams, J. E., 10 Finsbury Circus
Adlard, J. E., 22½ Bartholomew Close
Allingham, W., 25 Grosvenor Street
Allen, E., 11 Ave Maria Lane
Allen, E. G., 12 Tavistock Row (3 copies):—
Peabody Institute; Baltimore University of Toronto; Anon.
Anderson, M. S., 2 Gerrard Street, N.
Andrew, J., M.D., 22 Harley Street
Andrew, J., M.D., Everleigh House, Prince Arthur's Road, Hampstead
Aveling, J. H., M.D., 1 Upper Wimpole Street
Aveling, C. T., 12 Portland Place, Lower Clapton

Bain, J., 1 Haymarket, S.W.
Baker & Co., 4 Bond Court, Walbrook
Ballard, E., M.D., 12 Highbury Terrace, Islington
Bailliere & Co., King William Street
Barwell, Richard, 32 George Street
Bantock, G. G., M.D., 12 Granville Place, Portman Square
Barlow, Thomas, M.D., 10 Montague Street, Russell Square
Barclay, A. W., M.D., 23A Bruton Street
Barnes, R., M.D., 15 Harley Street
Barker, A. J., M.D., Ivy Lodge, Hornsey Road
Bartlett, J. H., 35 Ladbroke Gardens, Notting Hill
Barton, J. K., 88 Gloucester Road, Queen's Gate, S.W.
Bateman, H., 13 Canonbury Lane
Battye, R. F., 123 St. George's Road, S.W.
Baxter, E. B., M.D., 28 Weymouth Street, W.
Belgrave and Chelsea Medical Book Club
Bell, H. Royes, 44 Harley Street
Beale, L., M.D., F.R.S., 61 Grosvenor Street
Bennett, R., M.D., 22 Cavendish Square
Benington, R. C., Rose Bank, Coplestone Road, Peckham
Birkett, E. L., M.D., 48, Russell Square
Blake, E. T., M.D., 47 Seymour Street, Hyde Park
Bowman, W., F.R.S., 5 Clifford Street
Boulter, H. B., Library, St. Bartholomews
Bovill, E., 32 James Street, Buckingham Gate
Bristowe, J. S., M.D., 11 Old Burlington Street
Broadbent, W. H., M.D., 34 Seymour Street, Portman Square
Brookfield, J. Stores, M.D., 2 Devonshire Villas, Brondesbury, N.W.
Brown, Crichton, M.D., 7 Cumberland Terrace
Brown, Dyce, M.D., 29 Seymour Street
Brown, J. H., M.D., 55 Gordon Square
Brown, F. G., 16 Finsbury Circus
Brown, C. Gaye, M.D., 88 Sloane Street, Chelsea
Brown-Sequard, —, M.D., 44 Russell Square

Brunton, L., M.D., 50 Welbeck Street
Bryant, W. J., M.D., 23A Sussex Square, W.
Bryant, Thomas, 53 Upper Brook Street
Buchanan, G., M.D., 24 Nottingham Place
Burrows, Sir G., M.D., 18 Cavendish Square
Buzzard, T., M.D., 56 Grosvenor Street, W.

Carfrae, G. M., M.D., 4 Hertford Street, Mayfair
Chambers, Thomas, M.D., 64 Chester Square
Channer, H. O., M.D.
Charing Cross Hospital
Cheadle, W. B., M.D., 2 Hyde Park Place, Cumberland Gate
Chepmell, E. C., M.D.
Cheyne, R. R., 27 Nottingham Place
Chippendale, J., 16 Upper Phillimore Place, Kensington, S.W.
Chiara, —, M.D.
Chisholm, E., M.D.
Cholmeley, W., M.D., 63 Grosvenor Street
Church, W. S., M.D., 130 Harley Street
Clark, Andrew, M.D., 16 Cavendish Square, W.
Clapton, A., M.D., 10A St. Thomas Street, S.E.
Claremont, C., Milbrook House, Hampstead Road
Clifton, N. H., 20 Cross Street, Islington
Clover, J. T., 3 Cavendish Place
Collins, F., M.D., 7 Charter House Square, E.C.
Cooke, R. H., Church Street, Stoke Newington
Corner, F. M., Manor House, East India Road, Poplar
Cory, R., M.D., 14 Palace Road, Lambeth Road, S.E.
Coryn, W. J., M.D., 68 Acre Lane, Brixton Road
Couper, J., M.D., 80 Grosvenor Street
Cowell, G., 19 George Street, Hanover Square
Critchett, G., 21 Harley Street
Crocker, Radcliffe, M.D., 28 Welbeck Street
Croft, J., 61 Brook Street, Grosvenor Square
Crosby, T. B., M.D., 21 Gordon Square
Curgenven, J. B., 11 Craven Hill Gardens, Bayswater

Daniell, R. T., M.D., 20 Cathcart Road, West Brompton
Davidson, J., King's College Hospital
Davis, F. W., R.N.
Davies, H., M.D., 23 Finsbury Square, E.C.
Dawson, Yelverton, 4 Sydney Street, Fulham Road, S.W.
Dawson & Son, 121 Cannon Street (for Senores Medina Hermanos)
Dewar, J., 132 Sloane Street
Dickson, J. W., M.D., 23 The Grove, Hackney
Dingley, R., 7 Argyle Square
Dobell, H. B., M.D., 84 Harley Street
Dowdeswell, G., Windham Club, St. James's Square
Down, J. L. H., M.D., 39 Welbeck Street, W.

Drury, C. H., M.D., 3 Bucklersbury
Duckworth, Dyce, M.D., 11 Grafton Street, W.
Dudgeon, R. E., M.D., 53 Montague Square
Dunbar, James, 77 Ladbroke Grove, Kensington Park
Durham, A. E., 82 Brook Street, W.
Duncan, J., M.D., 8 Henrietta Street, Covent Garden
Duncan, J. Mathews, M.D., 71 Brook Street, W.
Duncan, H. M., 139 Buckingham Palace Road

Easton, J., M.D., 19 Norfolk Crescent, Hyde Park
Engall, Thomas, 15 Euston Square
Erichsen, J. E., 6 Cavendish Place
Evans, W. T., M.D., 21 Westbourne Villas, Harrow Road
Evans, T. C., 99 Camden Street

Fagge, C. H., M.D.. 11 St. Thomas Street
Fardon, E. A., Middlesex Hospital
Farr, G. F., M.D., 175 Kennington Road, S.E.
Fearnside, H., M.D., 49 Leinster Gardens, Hyde Park
Fitzgerald, W. A., 195 Loughborough Road, Brixton
Fenwick, S., M.D.. 29 Harley Street
Fleming, J. N., M.D., Champion Hill, Camberwell
Fletcher, J. C., M.D., 149 Camden Road
Ford, Edwin M., Avenue House, Peckham Rye
Fotherby, H. T., M.D., 3 Finsbury Square
Fox, Wilson, M.D., 67 Grosvenor Street
Fowler, G. F., 3 & 4 New Inn Yard, Shoreditch (2)
Francis, C. R., M.B., 1 Nelson Terrace, Clapham Common
Fuller, H. R., 19 Granville Place, Portman Square

Gannon, J. P.
Garden, A., M.D.
Garlick, W., M.D., 33 Great James Street, Bedford Row
Gibbs, H., M.D., 42 Colville Terrace, Bayswater
Gibson, J. R., 10 Russell Square
Giles, G., 11 North Terrace, Alexander Square, Brompton
Godrich, F., 140 Fulham Road, West Brompton
Goodhart, J. F., M.D., 27 Weymouth Street
Gowlland, P. Y., 34 Finsbury Square, E.C.
Graham, J., M.D., 29 Glo'ster Road, Regent's Park
Greenwood, A., 178 Cold Harbour Lane, Camberwell
Grindlay & Co., Parliament Street:—
Dr. J. Monteith
Dr. M'Connor
Guy, Thomas, M.D., 23 Auriol Road, Kensington
Gull, Sir W., M.D., Bart., 74 Brook Street
Gunn, R. Marcus, Royal London Ophthalmic Hospital, Moorfields

Hamilton & Co., Paternoster Row (6 copies)

Hall, De Haviland, M.D., Westminster Hospital
Habershon, S. O., M.D., 70 Brook Street, W.
Harris, Vincent, M.D., 23 Upper Berkeley Street, W.
Harris, S. C., Herne House, Ribblesdale Road, Hornsey
Hare, C. J., M.D., 57 Brook Street
Harling, R. D., M.D., 16 Seymour Street, W.
Harrison, H. F. E., M.D., 9 Park Villas, Shepherd's Bush
Hague, Samuel, 277 Southampton Street, Camberwell
Hawkins, C. H., F.R.S., 26 Grosvenor Street, W.
Heath, Christopher, 36 Cavendish Square
Henry, Alex., M.D., 57 Doughty Street
Herman, —, M.D., 20 Finsbury Square, E.C.
Hewitt, P. G., Chesterfield Street, Mayfair
Hewitt, Grailey, M.D., 36 Berkeley Square
Hope, W. M., 181 Piccadilly
Holmes, T., 18 Great Cumberland Place
Holman, W. H., M.D., 68 Adelaide Road
Hood, Peter, M.D., 23 Lower Seymour Street, W.
Hood, D. W. C., M.D., 43 Green Street, Park Lane
Hugman, W., 55 Guildford Street, Russell Square
Hunter, Bernard, 15 Grafton Street East
Hunterian Society, London Institution, Finsbury Circus
Hutchinson, J., 15 Cavendish Square
Hutton, R. J., M.D., 240 City Road, E.C.

Ironside, R. A., M.D., 8, Highbury New Park

Jackson, Hughlings, M.D., 3 Manchester Square
Johnson, W. B., 2 York Road, Lambeth
Johnson, G., M.D., 11 Savile Row
Jones, C. H., M.D., 49 Green Street, Park Lane
Jones, Sydney, M.B., 16 George Street, Hanover Square

Keep, C., Guy's Hospital
Kibbler, R. C., L.R.C.P., Granton House, King Edward's Road, South Hackney
Kimpton, R., 31 Wardour Street (2 copies)
Kimpton, H., High Holborn (3 copies)
King's College Library
King, H. S. & Co., Cornhill:—
- Moore, R. W., M.D.
- Reed, A. G.
- Ruttonjee, Hormusjee
- Miller, T. French, M.D.
- Jamieson, R. A., M.D.
- Myers, W. W.
- Berbice Society
- Chandra, R. C., M.D.
- Calcutta School Book Society

Chesnaye, Surgeon-Major
Menzies, J. A., M.D.
Henston, —, M.D.
Cameron, L., M.D.
Kisch, A., M.D., 46 Portsdown Road, Maida Vale
Knight, C. F., 139 St. John Street, Clerkenwell

Lansberg, P. von
Langmore, J. C., M.B., 20 Oxford Terrace
Lawrence, H. Cripps, L.R.C.P.L., 49 Oxford Terrace
Lawrence, J. E., East Hill, Wandsworth
Lawson, R., 20 Lansdowne Road, Notting Hill
Leadam, Ward, M.D., 80 Gloucester Terrace, Hyde Park
Lewis, H. K., 136 Gower Street, W.C.
Ligertwood, J., M.D., Royal Hospital, Chelsea
Lister, Prof., 12 Park Crescent, Regent's Park
Little, W. J., M.D., 18 Park Street
Loane, Joseph, Dock Street, Whitechapel
Lockhart, W., M.D., Park Villa, 67 Granville Park, Blackheath
Lockwood & Co., 7 Stationers' Hall Court:—
Columbus School for Feeble Minded, *Ohio*
Columbus Hospital for Insane, *Ohio*
Athens Hospital for Insane, *Ohio*
Cincinnati Hospital Library, *Ohio*
Murphy, J. A., M.D., *Cincinnati, Ohio*
Brown, W. T., M.D., *Cincinnati, Ohio*
Shepard, L. R., M.D., *Cincinnati, Ohio*
Parvin, T., M.D., *Indianapolis, Indiana*
Yandell, D. Y., M.D., *Louisville*
Meisse, J., M.D., *Chillicottie*
Stillwell, J. A., M.D., *Brownstown*
Seigler, J. A., M.D., *Brownsville, Indiana*
Ohio Eclectic Medical Association; and 2 others
London Hospital Library, Mile End
London School of Medicine for Women, Henrietta Street, Cavendish Square
Longman & Co., Paternoster Row:—
Alston, —, M.D.
Blake, —, M.D.
Garrison Medical Officers' Library
Gibralta Garrison Library
Tolmie, Surgeon-Major T. C. A., M.D.
Rosenburg, —, M.D.
Stokes, —, M.D., *Gibralta*
Gozel, —, M.D.
Piddington, W. R. (2 copies)
Lubbock, M., M.D., 6 Grosvenor Street
Low & Co., 188 Fleet Street:—
Brown, W., M.D., *Dunedin*

Bowker, C. S., M.D.
Ashurst, S., M.D.
Carter, C., M.D.
Burns, R., M.D.
Page, E. A., M.D.

Mackenzie, S., M.D., 26 Finsbury Square
Mackenzie, Morell, 19 Harley Street
Maclaren, A. C., M.D., 60 Harley Street
M'Cormac, W., 13 Harley Street
M'Farlane, W., M.D., 15 Lower Phillimore Place, W.
M'Keelmie, T. H., 60 Wimpole Street
Markwick, A., M.D., 1 Leinster Square
Mason, S., M.D., 44 Finsbury Circus
Meadowes, A., M.D., 27 George Street, Hanover Square
Medico-Chirurgical Society, Berners Street
Medical Society University College
Medical Book Society, 30 Queen Street, Cheapside
Medical Society of London, 11 Chandos Street, Cavendish Square
Meryon, E., M.D., 14 Clarges Street, Piccadilly
Mickley, G., M.B., M.A., St. Luke's Hospital
Middlesex Hospital Library, Berners Street
Miller, C. M., M.D., Claremont Villa, 86 Stoke Newington Road
Moline, Paul F., University College
Moxon, W., M.D., 6 Finsbury Circus
Mouat, F. J., M.D., 12 Durham Villas, Kensington, W.
Muir, J. C., L.R.C.P., 44 Cornwall Road, Westbourne Park
Muckerjee, S., M.D., 7 Crescent Place, Mornington Crescent
Murphy, S. F., 158 Camden Road
Myers, A. T., 1 St. George's Place, Hyde Park Corner

Nash, E., M.D., 123 Lansdowne Road, Notting Hill
Neatby, E. A., 2 Christchurch Road, Hampstead
Neatby, T., M.D., 29 Thurloe Road, Hampstead
Needham, J., M.D., 2 Westbury Gardens, Clapham Road
Nettleship, E., 4 Wimpole Street
North London Medical Book Society, 33 Parkhurst Road, Holloway, N.

O'Brien, B., M.D., 26 Somerleyton Road, Brixton
Oldham, H., M.D., 4 Cavendish Place

Paget, Sir J., F.R.S., 1 Harewood Place, Hanover Square
Palfrey, J., M.D., 29 Brook Street, Grosvenor Square
Panioty, John E., 2 St. Lawrence Road, Notting Hill
Paul, J. L., M.D., 43 Queensborough Terrace, W.
Peacock, T. B., M.D., 20 Finsbury Circus, E.C.
Peirce, R. King, 94 Addison Road, Kensington
Peirce, J. Channing, M.D., Manor House, Brixton Rise

Perkins, Houghton, 78 Mortimer Street, Cavendish Square
Pick, T. P., 13 South Eaton Place, Eaton Square
Poore, G. V., M.D., 30 Wimpole Street, W.
Portman Medical Book Club, 49 Seymour Street
Potter, J., M.D., 20 George Street, Hanover Square
Potts, W., M.D., 2 Albert Terrace, Regent's Park
Powell, R. D., M.D., 15 Henrietta Street, Cavendish Square
Power, H., M.B., 37A Great Cumberland Place, Hyde Park
Powdrell, J., 160 Euston Road
Priestley, W. O., M.D., 17 Hertford Street, Mayfair
Purnell, J. J., Woodlands, Streatham Hill

Quain, R., M.D., 67 Harley Street
Quain, R., F.R.S., 32 Cavendish Square

Ramskill, J. S., M.D., 5 St. Helen's Place, Bishopsgate
Ringer, Sydney, M.D., 15 Cavendish Place
Rivington, Walter, 22 Finsbury Square
Roberts, J. H., Hill Crest, Greenhill Road, Hampstead
Roberts, A., Hill Crest, Greenhill Road, Hampstead
Roberts, D. W., 56 Manchester Street
Roberts, F., M.D., 53 Harley Street
Robertson, G., Melbourne, and 17 Warwick Square (3 copies)
Roper, G., 6 West Street, Finsbury Circus
Ross, D. M., M.D., 54 Upper Berkeley Street, W.
Roth, M., M.D., 48 Wimpole Street

Sansom, A. E., M.D., 30 Devonshire Street, Portland Place
Saunders, W. S., M.D., 13 Queen Street, Cheapside
Savage, G. H., M.D., Bethlehem Royal Hospital, St. George's Road, Lambeth
Scott, J., M.D., 8 Chandos Street, Cavendish Square
Schmidt, A. E., M.D., 150 Bethnal Green Road
Sedgwick, L. W., M.D., 2 Gloucester Terrace, Hyde Park
Senior, Charles, Adelaide House, 22 Hilgrove Road, N.
Seager, Herbert W., St. Mary's Hospital
Sewell, C. B., M.D., 13 Fenchurch Street, E.C.
Seton, D. E., M.D., 12 Thurloe Place, S.W.
Skinner, W., 45 Lower Belgrave Street, S.W.
Shillitoe, B., 2 Frederick Place, Old Jewry
Sibley, S. W., 7 Harley Street
Silcock, A. Q., 5 Graham Road, Dalston
Simpkin & Co., Stationers' Hall Court:—
Van de Laan; and 3 others
Skelding, J., 16 Euston Square
Skeat, —, 10 King William Street
Smith, E. Noble, L.R.C.P.
Smith, Eustace, M.D., 5 George Street, Hanover Square
Smith, Protheroe, M.D., 42 Park Street

Smith, Fredk., 730 Old Kent Road
Smith, Walter, M.D., 2 Stanhope Terrace, Gloucester Gate
Smith, Gilbert, M.D., 68 Harley Street
Snell, E. G. C., M.D., 131 Green Street, Victoria Park
Sotheran, 136 Strand :—
  Haslar Hospital
  Plymouth Hospital
  Chatham Hospital
  Edulgee Musserwangee
South London Medical Reading Society, 148 Lambeth Road, S.E.
Squire, B., 24 Weymouth Street
St. Bartholomew's Hospital Library
St. Mary's Hospital Library
St. George's Hospital Library
Stevens, Felix, M.D., 13 High Street, Stoke Newington
Stevens, B. F., 4 Trafalgar Square (4 copies)
Stewart, W. E., 16 Harley Street
Stewart, A. P., M.D., 75 Grosvenor Street
Stewart, H. C., M.D., 22 North Bank, Regent's Park
Stoke Newington Medical Society, 57 Darnley Road, Hackney, E.
Surgeons, Royal College of, Lincoln's Inn Fields
Symonds, J., M.D., 79 Amhurst Road, Hackney

Tamburini, —, M.D.
Tayloe, E., South Lodge, Clapham Common
Tegart, E., 49 Jermyn Street, S.W.
Tenison, E. T. R., M.D., 9 Keith Terrace, Shepherd's Bush
Thacker & Co., Newgate Street :—
  Charles, —, M.D.
  Manhook, —, M.D.
  Udoz, Chand Dutt, M.D.
  Benode, K. Bose, M.D.
Thorowgood, J. C., M.D., 61 Welbeck Street
Thompson, A., M.D., 10 Delamere Street, W.
Thompson, Sir H., 35 Wimpole Street
Thyne, T., M.D., 140 Minories, E.C.
Toulmin, F., Upper Clapton, N.E.
Travers, W., M.D., 2 Phillimore Gardens, S.W.
Turner, F. Charelwood, M.D., M.A., 15 Finsbury Square
Tweedy, J., M.D., 18 Harley Street
Twynam, G. E., 18 Blandford Square, W.
Trübner & Co., 57 Ludgate Hill (11 copies)

Vaillant, Edward, 85 George Street, Portman Square
Venning, Edgecombe, 87 Sloane Street, W.
Vereker-Bindon, —, M.D., 2 Elm Villas, Willesden Lane, Kilburn

Wake, —, M.D., 2 Cathcart Hill, Holloway, N.
Waring, E. J., M.D., 49 Clifton Gardens, Maida Vale

Waggett, J., M.D., 40 Ladbroke Grove, Kensington Park Gardens
Walker, J. P., M.D., 30 Bedford Square
Waller, A., 14 Gibson Square, Islington
Warner, Percy, 4 Merrick Square, Borough
Warren, E. C., M.D., 99 Albion Road, Dalston
Watkins, S. C., Poplar Hospital for Accidents, East India Road, Poplar
Watkins, E., M.D,, 61 Guildford Street
Watkins, C. J., 27 Mornington Crescent
Watney, Herbert, M.D.. 1 Wilton Crescent, S.W.
Watson, Sir T., M.D., Bart., 16 Henrietta Street, Cavendish Square
Watson, J., M.D., 6 Southampton Street, Bloomsbury
Waylen, A., M.D.
Webb, F., M.D., 113 Maida Vale
Weber, H., M.D., 10 Grosvenor Street
Weber, F., M.D., 44 Green Street, Grosvenor Square
Webster & Larking, Piccadilly
Wells, T. Spencer, 3 Upper Grosvenor Street
Wesley, W., Strand
Weston, Philip, M.D., 391 City Road
Whaley, J. C., Prospect Place, Kilburn
Whitmore, W. T., 7 Arlington Street, S.W.
Wigg, A. E., University College Hospital
Williams & Norgate, Henrietta Street
Williams, Theodore, M.D., 47 Upper Brook Street
Williams, J., M.D., 44 Mildmay Park, N.
Williams, Dawson
Wilkin, J. C., 15 Hyde Park Street
Wilks, S., M.D., 72 Grosvenor Street
Wilson, Erasmus, F.R.S., 17 Henrietta Street, Cavendish Square
Worley, W. C., 43 De Beauvoir Road, W.
Worsley-Benison, H. W. S., 25 Grange Road, Canonbury, N.

Loughborough *see* Leicester
LOSTWITHIEL .... Row, C., M.D.
Loughor *see* Llanelly
LOUTH .... Faussett, F.
LOWESTOFT .... CLUBBE, W. H., *Loc. Sec.*
Ray, J.
Worthington, F. S.
LOWER TULSE HILL ...... Stowers, J. Herbert, M.D.
LUTON
LUTTERWORTH .... Busgard, M., M.D.
LYMINGTON .... Hill, W. R., M.D.
Lyndhurst *see* Southampton
LYNN .... WOODWARD, E., L.R.C.P., *Loc. Sec.*
West Norfolk and Lynn Hospital
Parry, G., *Docking*
Webster, W.

MABLETHORPE
MACCLESFIELD
MADELEY
MAIDENHEAD
MAIDSTONE .... Plomley, I. F.
Ground, E.
MALPAS .... Parker, R., M.D.
MALTON .... COLBY, W. T., M.D., *Loc. Sec.*
MALVERN .... WEST, W. C., M.D., *Loc. Sec.*
Dawson, W. H.
Rowland, H. M., M.D.
Weir, A., M.D., *St. Mungho's*
MANCHESTER .... PEATSON, J. CHADWICK, M.D., 25 Mount Street, Peter Street, *Loc. Sec.*
Armstrong, J., *Harpurhey*
Blackley, C. H., *Trafford*
Borchardt, L., M.D.
Bradbury, J. O., *Salford*
Bradshaw, J. D., *Bowdon*
Brown, H., M.D., *Heaton Mersey*
Buckley, S., F.R.C.S.
Child, W. L., *Prestwich*
Clarke, A. C. S. W., *Salford*
Clarke, R., M.D., *Farnsworth*
Coveney, J. H., *Prestwich*
Crompton, S., M.D., *Cheetham Hill*
Earl, J., *Cheetham Hill*
Fletcher, J. S., M.D., *Higher Broughton*
Fox, R. D.
Gornall, R. H., *Newton Heath*
Heathcote, G.
Hewson, R. W., *Cheadle*
Heslop, R.
Hodgkinson, Alex., M.D.
Humphrey, H., M.D., *Eccles*
Ilderton, F., *Fairfield*
Irwin, J., M.D.
Jones, T., M.D.
Kennedy, J.
Lancashire, J., *Stand*
Leach, D. J.
Library of Royal Infirmary
Library of Medical Society
Land, E.
Mallett, W. J., M.D.
Morgan, J. E., M.D., *Eccles*
Mould, G. W., M.D.
Mules, P. H., M.D., *Bowdon*
Murphy, C. O.

Manchester, *continued* ... Nesfield, S., M.D.
Paton, Robert
Phipps, G. C., M.D.
Peirce, F. M., M.D.
Radford, T., M.D., *Higher Broughton*
Ransome, M. A., M.B., *Bowdon*
Reed, G., M.D.
Renshaw, J., M.D., *Stretford*
Renshaw, S. H., M.D., *Sale*
Roberts, D. L., M.D.
Roberts, W., M.D.
Rodger, R., M.D.
Roe, R. E. H., *Patricroft*
Ross, James
Simpson, H., M.D.
Smale, H. C.
Smart, R. B.
Southam, F.
Stephens, J.
Skinner, C. G. L., *Harpurhey*
Simon, R. M., M.D.
Stone, J., M.D.
Walmsley, Francis H.
Walter, W., M.D.
Wattie, Alex., M.A., M.B.
Westmorland, J., *Cheetham*
White, J. A., *Pendleton*
Williams, W. J., M.D.
Withington, G. H., *Kersall*

March *see* St. Ives (Huntingdon)
Margate .... Rowe, T. S., M.D.
Market Deeping .... Forster, H. J.
Market Drayton
Market Overton .... Roe, W.
Market Rasen .... Taplin, B. D., *Binbrook*
Markfield *see* Leicester
Marlborough ....
Marple .... Bailey, J. Johnson, M.D., *Loc. Sec.*
Martock
Maryport .... Spurgin, W. H.
Crerar, J., M.R.C.P.
Melksham .... King, J. R.
Keir, W. Ingram
Melton Mowbray
Merthyr Tydvil .... Biddle, Cornelius, L.R.C.P. Lond., *Loc. Sec.*
Mickleover................ Lindsay, J. Murray, M.D.
Middlesborough-on-Tees Hedley, J., *Loc. Sec.*
Glen, J., M.D.

MIDDLESBORO'-ON-TEES, *con.* ...... Ketchen, W., M.D.
White, W.
Williams, W. J. T., M.D.

Milbrook *see* Southampton
MITCHAM .... Marshall, Ed.
Mold *see* Chester
MONMOUTH
MORETON-IN-THE-MARSH
MORPETH .... Logie, Jas., M.D.
MOSELEY .... Blake, G. F.
Shaw, Oliver C.

Nafferton *see* Bridlington
NEEDHAM MARKET
Neston *see* Chester
NEWARK-ON-TRENT .... APPLEBY, F. H., *Loc. Sec.*
Luscombe, W. E., *North Collingham*
Hallowes, W. B.
NEWBURY .... BUNNY, JOSH., M.D., *Loc. Sec.*
Palmer, Montagu H. C.
Hickman, Richard
NEWCASTLE-UNDER-LYNE
NEWCASTLE-ON-TYNE ...... OLIVER, THOS., M.D., *Loc. Sec.*
Armstrong, L.
Arnison, W. C., M.D.
Atkinson, J. J., *Wylam-on-Tyne*
Barkus, B., M.D., *Gateshead*
Beatly, T. C., jun., *Seaham Harbour*
Benson, T., *Stanley*
Bowman, Hugh Torrington, M.A., M.B.
Brown, W. J., M.B.
Callcott, J. T., *Sedgefield*
Crisp, J. L., *South Shields*
Downie, G., *Chester-le-Street*
Gateshead Medical Society
Gibb, C. J., M.D.
Gibson, C., M.D.
Heath, G. Y., M.D.
Houseman, J., M.D.
Hume, G. H., M.D.
Jackson, D., M.D.
Kennedy, W. A.
Library of Newcastle Infirmary
Macaulay, J., M.D.
Matthews, J., M.D., *Tynemouth*
M'Coull, G., L.R.C.P. Ed., *Orington-on-Tyne*
Murray, W., M.D.
Murray, J. C., M.D.

NEWCASTLE-ON-TYNE, *con.*	Nesham, T. C., M.D.
	Rayne, S. W.
	Renton, W., M.D., *Shotley Bridge*
	Service, John, *West Bolden*
	Smith, J. W., M.D., *Ryton-on-Tyne*
	Smith, J.
	Stainthorpe, G. F.
	Sutherland, W., *Capheaton*
	Wilkinson, Auburn
	Wilson, R. H., M.D., *Gateshead*
	Wilson, W. T., M.D.
	Wilson, J., M.D., *Lanchester*
NEWENT	Smelt, F. H., L.R.C.P.E.
NEWICK, UCKFIELD	Graveley, Richard
NEWMARKET	
NEWPORT (Monmouthshire)	MORGAN, W. W., M.D., *Loc. Sec.*
	Limberry, Thos.
	Ready, W. J. Markham
NEWPORT (Pembrokeshire)	Havard, D., M.D.
NEWTON ABBOTT	
NORHAM-ON-TWEED	Paxton, J., M.D.
NORTHAMPTON	EVANS, C. J., *Loc. Sec.*
	Banks, P. H., *Riseley*
	Busgard, —, M.D.
	Clarke, W. W., M.B., *Wellingborough*
	De Denne, T. V.
	Infirmary Library
	Moxon, W.
North Curry *see* Taunton	
NORTH SHIELDS	PEART, R., M.D., *Loc. Sec.*
	Bourne, W., M.D.
	Stephens, Thos.
	Turnbull, T. J.
NORWICH	ROBINSON, H. S., *Loc. Sec.*
	Eade, P., M.D.
	Manby, F., *Reedham*
	Medico-Chirurgical Society
	Taylor, Hugh, *Cottishall*
NOTTINGHAM.................	RANSOME, W. H., M.D., *Loc. Sec.*
	Beddard, J., M.B.
	Brookhouse, J. O., M.D.
	General Hospital Library
	Howitt, F., M.D.
	Phillimore, W. P., M.D.
	Terrewest, Miss
	Wight, Joseph
	Wright, Thomas, M.D.
NUNEATON	Hammond, W., M.D.

Oaken	Hawthorn, F. J.
Odiham	M'Intyre, J., M.D., *Loc. Sec.*
Oldham	Platt, Thos., *Loc. Sec.*
	Fawsitt, T.
	M'Gowan, S. A., M.D.
	Thomson, G., M.D.
Ormskirk.....................	Wickham, H., *Rufford*
Ossett *see* Leeds	
Oswestry	
Oswaldtwistle	Townley, A. T.
Otley	Ritchie, Thos.
	Scott, Thos.
Ottery St. Mary	Hemsted, A., L.R.C.P.
Over *see* St. Ives	
Over Darwen...............	Wraith, J. H.
Oxford	Winkfield, A., F.R.C.S., *Loc. Sec.*
	Ackland, W. H., M.D.
	Chapman, E.
	Freeborn, R. F., M.D.
	Jackson, R., M.D.
	Morgan, W. L.
	Medico-Chirurgical Society
	Radcliffe Library
	Oxley, A. R.
	Symonds, F.
	Thompson, Harold
	Thompson, W. A.
	Ward, J. B., M.D.
Painswick	Sampson, H. M.
Pendleton *see* Manchester	
Peniston *see* Sheffield	
Pembroke Dock............	Stamper, J. F., M.D.
Penge	Peck, R. H.
	Watson, G. S.
Penrith	Wickham, J., M.D.
	Jackson, T., M.D.
Pen-y-groes	Roberts, Evan
Penzance.....................	Montgomery, J., M.D., *Loc. Sec.*
	Boase, F.
	Grenfell, H.
	Bennett, W. S.
	Tresize, W. R., *St. Just*
	Berry, W. A.
Peterborough	Cane, Leonard, M.D., *Loc. Sec.*
	Walker, T. J., M.D.
	Thomson, W., M.D.
Pewsey *see* Devizes	
Plaistow.....................	Kennedy, A. E.

Plumstead	Smith, Henry
Plymouth	Whipple, Connell, *Loc. Sec.*
	Bazeley, W.
	Clay, W. H., M.D.
	Eccles, G. H., M.D.
	Elliott, R. M., *Kingsbridge*
	Harper, T.
	May, G. H. T.
	Plymouth Book Society
	Meeres, E. E., M.D.
	Square, W. J.
	Swain, P. W., *Devonport*
	Willis, R., *Horrabridge*
Pontypool	Mason, S. B.
Poole	
Portland	Trevan, M., R.N.
Portsmouth	Ford, A. Vernon, L.R.Q.C.P.
Portsea	
Portmadoc	Griffiths, S., M.D.
	Morris, W. Jones, L.R.C.P.E.
Preston	Allen, R., *Loc. Sec.*
	Arminson, W. B., M.D.
	Berry, J., *Leyland*
	Christison, J., M.D.
	Hall, J.
	Hammond, J. H., M.D.
	Heslop, R. C., M.D.
	Gilbertson, J. B., M.D.
	Rigby, J.
Prescot	
Pulborough	Taylor, W. E., M.D.
Queen Camel	Stovin, C. F., M.D.
Rainhill	Wigglesworth, J., M.B.
Ramsgate	Walford, E., *Loc. Sec.*
	Glanville, F. F.
Ravensthorpe	Ramsden, W. T.
Rawtenstall	Edward, A. A., M.D.
Reading	Walford, T. L., *Loc. Sec.*
	Book Society
	May, G.
	Armstrong, —, M.D.
	Graham, C. R.
	Hayes, H. R., *Basingstoke*
	Lowsley, Oded
	Workman, J. W.
	Young, W. B.
Redland	Webster, Thomas

REDRUTH	Hichens, J. S.
	Michell, G. A.
REIGATE	WALTERS, J., M.D., *Loc. Sec.*
	Blake, E., M.D.
	Holman, C., M.D.
	Smith, T. P., M.B.
RETFORD	
RHAYADER	Richardson, R.
RICHMOND (Surrey).........	Fenn, E., M.D.
RICHMOND (Yorks)	Bowes, R.
RIPON	
RIPPINGALE	Adams, G. N., M.D.
ROCHESTER	Langston, J., *Strood*
	White, C. J., M.D., *Snodland*
	Tribe, H. H., *Chatham*
ROCHDALE	POOLEY, R. M., *Loc. Sec.*
ROCHFORD	KING, THOMAS, M.D., *Loc. Sec.*
Rockferry *see* Birkenhead	
ROMFORD	
ROTHERHAM	Foote, H. D., M.D., M.R.C.S.
	Blythman, C. S., M.B., M.R.C.S., *Swinton*
	Burman, W. M., L.R.C.P., M.R.C.S., L.S.A., *Wath-on-Deane*
	Brett, J., L.R.C.P. Lond., M.R.C.S. England
	Clark, W., M.D., L.R.C.S.E., *Wentworth*
	Gowan, Charles, M.D., *Auston*
	Jones, W. M., M.R.C.S., L.S.A.
	Knight, H. J., M.R.C.S.
	Lyth, J. B., M.R.C.S.E., L.R.C.P. Ed.
	Oxley, W., M.R.C.S., L.S.A.
	Saville, W., M.R.C.S., L.S.A.
	Smith, W. J., *Parkgate*
ROTHWELL	More, J., M.D.
Royston *see* Cambridge	
RUDDINGTON..................	Hall, J., M.D.
RUGBY	Dukes, Clement, M.B.
	Simpson, Herbert, M.D.
RYDE	Buck, T. A., M.D.
SAFFRON WALDEN	STEAR, H., *Loc. Sec.*
ST. ALBANS..................	Prior, R. H., M.D.
ST. CLEARS..................	Jones, V., Ll.
ST. GERMANS	Kerswell, J. B.
ST. HELENS	TWYFORD, E. P., M.D., *Loc. Sec.*
	Gaskell, R. A.
	Jamieson, A., M.D.
	Martin, J. H.

St. Ives (Cornwall)	Joll, Boyd B., M.B.
St. Ives (Huntingdon)...	Grove, W. R., M.D., *Loc. Sec.*
	Dixon, Rev. W., *Over*
	Osborne, Harold Rochester
	Easby, —, M.D., *March*
St. Just *see* Penzance	
St. Tudye	Pearse, W.
St. Leonards *see* Hastings	
Salford *see* Manchester	
Salisbury	Wilkes, W. D., *Loc. Sec.*
	Darke, F. R. P.
	Kelland, James
	Gordon, James, M.D.
	Lee, F. F.
Sandon........................	Tylecote, J. H., M.D.
Sawbridgeworth............	Brickwell, J.
Scarborough	Cooke, R. B., *Loc. Sec.*
	Murray, Ivor, M.D.
	Wandby, W.
Seacombe *see* Birkenhead	
Sedburgh	Green, T. B.
Sedgefield.................	Smith, R., M.D.
Sevenoaks	Alliott, A. J., M.B.
Shaftesbury (Dorset)	Wilkinson, A., M.D.
Shanklin	
Sheepshed *see* Leicester	
Sheerness	Swales, E.
Sheffield	Martin de Bartholomé, M., M.D., *Loc. Sec.*
	Banham, H. F., M.D.
	Barber, Jonathan
	Benson, John
	Booth, W. H.
	Branson, F., M.D.
	Drew, S., M.D., *Chapeltown*
	Favell, W. F.
	Gleadall, J.
	Gwynn, C. N.
	Hardwicke, J., M.D.
	Hardwicke, H. J., M.D.
	Hawthorn, H. J., *Ecclesfield*
	Jackson, A., M.R.C.S.
	Jones, J. T., *Eckington*
	Keeling, J. H., M.D.
	Martin, John W., *Dronfield*
	Porter, W. S.
	Roberts, J. S., M.D.
	Robinson, G., M.R.C.S.
	Sheffield Medical Book Society

SHEFFIELD, *continued*	Smith, R. J. Pye
	Taylor, G. Stopford
	Thomas, W. R., L.R.Q.C.P.
	Watson, T. H., M.B.
SHERBORNE	Williams, W. H., jun.
Shotley Bridge *see* Newcastle-on-Tyne	
SHREWSBURY	ANDREW, E., M.D., *Loc. Sec.*
	Eddowes, A.
	Edwards, H. N.
	Taylor, H. Coupland, M.D.
	Whitwell, F.
SIDMOUTH	
SLEAFORD	
SLOUGH	
SOUTHAMPTON	GRIFFIN, R. W., M.D., *Loc. Sec.*
	Bencraft, H.
	Dayman, H., *Milbrook*
	Nunn, G. R., *Lyndhurst*
	Oliver, J.
	Royal Victoria Hospital
	Southampton Medical Society
	Sims, W., M.D.
	Trend, T. W., M.D.
	Viant, H., L.R.C.P., *Totton*
	Ward, Thomas
SOUTHEND	
SOUTHPORT	Mort, W., M.D.
	Barron, A.
SOUTH SHIELDS	FRAIN, Jos., M.D., *Loc. Sec.*
	Armstrong, I. F., M.D.
	Bradley, W. M., M.D., *Jarrow*
	Crease, J. R.
	Hewitson, W.
	Robson, J.
	Robson, Adam
SOUTHSEA.....................	AXFORD, W. H., M.D., *Loc. Sec.*
	Manley, —, M.D.
	Maybury, L., M.D.
	Miller, J. W. Moore, M.D.
	Turner, G.
SPALDING	MORRIS, E., M.D., *Loc. Sec.*
	Vise, Ambrose Blythe, *Holbeach*
	Swan, R. Jocelyn, *Gosberton*
Spilsby *see* Boston	
STAFFORD......................	(Vacant) *Loc. Sec.*
	Blackford, J. C., M.D., *Cannock*
	Weston, E. J.
	Wynne, J. K., *Eccleshall*
STAINES	

Staleybridge	Booth, T. C.
Stamford....................	Newman, W., M.D., F.R.C.S., *Loc. Sec.*
	Medical Book Society
	Heward, J. M.
Stamford Bridge	Wright, F.
Stand *see* Manchester	
Stapleton	Levinge, E. G., M.D.
Stockport	Ball, J. A., *Heaton Norris, Loc. Sec.*
	Bagley, Samuel, *Hazel Grove*
	Bird, J. D., M.B., *Heaton Chapel*
	Downs, G., M.D.
	Greenhalgh, T., M.D., *Heaton Norris*
	Jordan, F. M., M.D., *Heaton Chapel*
	Massey, T.
	Turner, G., M.D.
	Whitehead, G. M.
Stockton-on-Tees	Oliver, W. H., *Loc. Sec.*
	Foss, R.
	Hind, H. F.
	Trotter, A. E. H.
	Young, C., M.D., *Yarm*
Stoke-upon-Trent	Johnson, Samuel, C., M.D., *Loc. Sec.*
	Arlidge, T. T., M.D.
Stourbridge	Freer, A., *Loc. Sec.*
	Ashmead, C., L.R.C.P., *Brierley Hill*
	Chapman, G., *Brierley Hill*
	Ker, Hugh R., *Cradley Heath*
	Oates, J. P.
	Thomson, Wesley, M.D., *Cradley*
Stratford-on-Avon	Nason, J. J., M.B., *Loc. Sec.*
Strood *see* Rochester	
Stroud........................	Cubitt, G. R.
Sudbury	Holden, J. S., M.D.
Sunbury-on-Thames *see* Surbiton	
Sunderland	Douglas, M., *Loc. Sec.*
	Bernard, G., M.D., *Silksworth*
	Brady, Prof. G.
	Broadbent, S. W.
	Dixon, W. H., M.D.
	Horan, J.
	Maling, E. A.
	Medical Society
	Morgan, G. B.
	Smith, Ayre, M.D.
	Watson, J., *South Hetton*
	Waterston, J.
	Welford, G.
	Wilson, J.
Surbiton	Kershaw, W. W., M.D., *Loc. Sec.*

SURBITON, *continued*	Coleman, M. T.
	Gunther, T. M., M.D.
	Izod, Charles, *Esher*
	Jones, Price, M.D.
	Kingsford, Edward, *Sunbury*
	Mott, Charles
	Tindall, W. R., M.D., *Hampton*
	Trouncer, J. H., M.D.
	Wyman, W. S., M.D., *Putney*
SUTTON	
SWANSEA	GRIFFITHS, T. D., M.B., *Loc. Sec.*
	Evans, J., L.R.C.P. Ed.
	Latimer, H. A.
	Mowat, G., M.D.
	Thomas, Jabez, M.D.
SWINDON	SWINHOE, G. M., L.R.C.P., *Loc. Sec.*
	Maclean, J. Campbell, M.B.
SYDENHAM	Erskine, W.
	Wilkinson, F. E., M.D.
TALARVOR....................	Roberts, J., M.D.
TAMWORTH	Joy, J. Holmes, M.D.
Tarvin *see* Chester	
TAUNTON	LIDDON, W., M.D., *Loc. Sec.*
	Farrant, S.
	Kinglake, H., M.D.
	Liddon, E., M.D.
	Olivey, H. P., *North Curry*
	Penny, H. S.
	Rigden, G. W.
	West, R. H., M.A.
TAVISTOCK	Crichton, —, M.D.
TEIGNMOUTH	Lake, W. C., M.D.
	Magrath, J. A., M.D.
TENTERDEN	Saunders, E.
TETBURY	
THAMES DITTON	
THETFORD	MINNS, P., M.D., *Loc. Sec.*
THAXTED	
THIRSK........................	Ryot, W. H., M.D.
Thorne and Tickhill *see* Doncaster	
TISBURY	Lindsay, R., M.D.
Tolleshunt D'Arcy *see* Colchester	
TOOTING	Ward, F. H.
TORQUAY	Cash, A. M., M.D.
	Heath, R., M.D., *Ellington*
TOTTENHAM	MAY, E. H., M.D., *Loc. Sec.*
	Cathcart, S.
	Cresswell, J., *Winchmore Hill*

Tottenham, *continued*......	Hutton, E. R.
Tonbridge	Ievers, Eyre, M.D.
Towcester	Weir, J. B., M.D.
	Evans, Arthur G.
Tredegar	O'Rorke, C., M.D.
Tring	Pope, E., M.D.
Trowbridge..................	Tayler, G. C.
Truro	Sharp, E., *Loc. Sec.*
	Leverton, H. Spry
	Royal Cornwall Infirmary Library
Tunbridge Wells	Barry, J. Milner, M.D., *Loc. Sec.*
	Chadwick, C., M.D.
	Manser, F.
	Johnson, J., M.D.
	Stamford, W.
	Ranking, J. E., M.B.
	Wallis, W., jun., *Hartfield*
	Wardell, J. R., M.D.
Tynemouth *see* Newcastle-upon-Tyne	
Ulverstone	
Upper Gornal	Walker, T. A.
Upper Norwood	
Uppingham	
Upton-on-Severn	Braddon, C.
Uxbridge....................	Macnamara, G. H., *Loc. Sec.*
Ventnor	
Wakefield	Wood, F. H., *Loc. Sec.*
	Clark, H.
	Holdsworth, S. R.
	Kemp, B.
	Kemp, B., jun.
	Major, —, M.D.
	Slatter, W. A.
	Walker, Thomas
	Wiseman, J. G., M.D., *Ossett*
Wallingford	Barrett, C. A., *Loc. Sec.*
	Barron, J., M.D.
	Greene, Walter, L.R.C.P.
Walsall	
Walton-on-Thames	
Warminster	
Warrington	Gornall, J. H., *Loc. Sec.*
Waterbeach	Grubb, J. S., M.D.
Watford	
Wath-on-Deane *see* Rotherham	
Wednesbury *see* West Bromwich	

WELLINGTON (Somerset)...	Edwards, Walter, M.D.
WELLS	FRENCH, J. G., *Loc. Sec.*
WEOBLEY	Applebe, E. A., M.D.
WEM	Wilson, J. G., M.D.
WEST BROMWICH	MANLEY, J., *Loc. Sec.*
	Evans, A. P.
	Garman, W. C., *Wednesbury*
	Sutcliff, H.
	Underhill, F. W.
	Underhill, T., *Great Bridge*
	Underhill, W., *Tipton*
WESTGATE-ON-SEA	Flint, Arthur, F.R.C.P.
WESTON-SUPER-MARE	ALFORD, R., *Loc. Sec.*
	Clark, —, M.D.
	Martin, Edward Fuller
	M'Clure, T. W., M.D., *Worle*
	Roxburgh, Robert
	Wickstead, F. W. S.
WEYMOUTH	
WEYBRIDGE	Graham, A. R., M.D.
WHITBY	YEOMAN, J., M.D., *Loc. Sec.*
	Lavenick, J. V., L.R.C.P. Lond.
	Mead, E. P., M.D.
WHITEHAVEN	I'ANSON, J. F., M.D., *Loc. Sec.*
	Braithwaite, S., *Egremont*
	Calderwood, G., M.D., *Egremont*
	Dickson, J., M.D.
	Henry, E. W., M.D.
	Horan, P. C., M.D.
	Parker, C., M.D., *Gosforth*
	Speirs, W., *Cleator*
WIGAN	TATHAM, G. G., M.D., *Loc. Sec.*
	Gallibrand, W., M.D.
WILLESDEN	
WIMBORNE	
WINCHESTER	BUTLER, F. J., M.D., *Loc. Sec.*
	Forder, Thomas
	Hants County Hospital Library
WINDERMERE	Hamilton, A., M.D.
Winchmore Hill *see* Tottenham	
WINDSOR	Ellison, J., M.D.
	Fairbank, Thomas, M.D.
	Harper, J. P., M.D.
WISBEACH	Lithgow, R. A. D., M.D.
WITNEY	BATT, A., M.D., *Loc. Sec.*
WOLVERHAMPTON	JACKSON, V., *Loc. Sec.*
	Cooke, J. B., M.B., *Tettenhall*
	Fraser, J., M.D.
	Newnham, C. A.

WOLVERHAMPTON, *continued*	M'Munn, C.A., M.D.
	Bell Medical Library
*WOOLWICH	MASON, R., *Loc. Sec.*
	Butler, W. H., L.R.C.P.
WORCESTER	Worcester Medical Society
WORKINGTON	Hight, J., M.D.
	M'Kerrow, G., M.B.
WORKSOP	Newton, Isaac
	O'Connor, D. M.
WORTHING	HARRIS, W. J., *Loc. Sec.*
WOTTON-UNDER-EDGE	Forty, D. H.
WREXHAM	DAVIES, E., M.D., *Loc. Sec.*
	Dickenson, J.
	Williams, E., M.D.
Wylam *see* Newcastle-on-Tyne	
YARMOUTH (Norfolk)	PALMER, C., *Loc. Sec.*
	Aldred, C. C., M.D.
	Mayo, A. C.
	Mitchell, Alex., M.D.
	Helston, —, M.D.
YARMOUTH (Isle of Wight)	Hollis, C. W., M.D.
YORK	SHANN, G., M.D., *Loc. Sec.*
	Ball, Alfred
	Draper, W.
	Dunhill, C. H., M.D.
	Gill, H. C., *Clifton*
	Hines, C. H.
	Hood, W.
	Jalland, W. H.
	Matterson, W., M.D.
	North, S. W.
	Ramsey, I., M.D.
	Read, W.
	Swanson, G. J., M.D.
	Williams, T. M.

SCOTLAND.

ABERDEEN	WIGHT, J., M.D., *Loc. Sec.*
	Adam, J.
	Davidson, A. D., M.D.
	Frazer, A., M.D.
	Garden, R. J., M.D.
	Jackson, H., M.D.
	M'Hardy, J. D., *Banchory*
	M'Robbie, J. S.
	Medico Chirurgical Society

ABERDEEN, *continued*	Mortimer, W., *Turriff*
	Ogston, F., M.D.
	Reith, Archd., M.D.
	Robb, J., M.D.
	Smith-Shand, W. F. J., M.D.
	University of Aberdeen
	Wallace, A., M.D., *Turriff*
	Ogilvie, Will., J. C., M.D.
	Willock, Richmond, M.D.
AIRDRIE	Arthur, H., M.D.
	Rankin, P., M.D.
ALEXANDRIA	M'Lelland, A., M.D.
ALLOA	Kirkwood, J., M.D.
	Wilson, R., M.D.
ANNAN	Drummond, T., M.D.
ARBROATH.....................	Walker, J. H., M.D., *Friskheim*
AUCHTERMUCHTY	Troup, F., M.D.
AYR	M'KERROW, G., M.D., *Loc. Sec.*
	Aitken, J., M.D., *Drongan*
	Dobbie, R., M.D.
	Highet, K., M.D., *Dalmellington*
	Riddall, —, M.D.
	M'Gill, J. F., M.D., *Annbank*
	Watt, J. R., M.D.
	Moore, —, M.D.
BANFF	Barclay, J., M.D.
	Manson, A. J.
Bannockburn *see* Stirling	
BARRHEAD	Corbett, R.
BERWICK-ON-TWEED.........	Fraser, T., M.D.
Bridge of Earn *see* Perth	
BUCKIE.......................	Duquid, W. R., M.D.
CAMPBELTOWN	Gibson, W., M.D.
	Cunningham, J., M.B.
CASTLE DOUGLAS...........	Lorraine, W., M.D.
CATRINE	Sloan, D., M.D.
COLDINGHAM-BY-AYRTON ...	M'Dougal, J. M., M.D.
COLDSTREAM.................	Turnbull, M. J., M.D.
CUMNOCK	Lawrence, J., M.D.
	Herbertson, R. G., L.R.C.P.
DALMALLY	M'Nicol, H.
DREGHORN	Hunter, J., M.D.
DUFFTOWN	Innes, J. A., M.D.
DUMFRIES	M'Culloch, J. M., M.D.
DUNDONALD	Alexander, W., M.D.
DUNFERMLINE	Henry, W., M.B.

DUNFERMLINE	Henry, W., M.B.
DUNSE	Stuart, J. A. E., M.D.
	M'Watt, J.
DUNDEE	Dewar, J. A., M.D., *Arbroath*
	Paton, D., M.D., *Carnoustie*
	Rorie, J., M.D.
	Sinclair, Robert, M.D.
	Wemys, J. W., M.D., *Broughty Ferry*
EDINBURGH	HUSBAND, W., M.D., *Loc. Sec.*
	Adams, Josh.
	Anderson, T., M.D., *Rosewell*
	Andson, W., M.D.
	Balfour, G., M.D.
	Balfour, Thos., M.D.
	Balfour, A., M.D., *Portobello*
	Berry, G. A., M.D.
	Bishop, J.
	Bose, C. C.
	Black, A.
	Beck, J. H. M., M.D.
	Bruce, R., M.D.
	Bryce, W., M.D.
	Brackenridge, D. J.
	Bramwell, Byron, M.D.
	Brown, J. Macdonald
	Bruce, Alex., M.D.
	Burns, J., M.D.
	Cadell, F.
	Cappie, J., M.D.
	Clouston, T. S., M.D.
	Craig, A., M.D., *Pathhead, Ford*
	Dickson, G., M.D.
	Dickson, A., M.D.
	Duncan, J., M.D.
	Duncanson, J. Kirk, M.D.
	Dunsmure, J., M.D.
	Falconer, John, M.D., *Lasswade*
	Fraser, Prof. T.
	Furley, R. C.
	Gordon, J., M.D., *East Linton*
	Gordon, P., M.D., *Juniper Green*
	Haldane, Rutherford, M.D.
	Hirschfeld, —, M.D.
	Jamieson, J., M.D.
	Jordan, Gregory P.
	Keillor, A., M.D.
	Keith, J. S., M.D.
	Kerr, W. W.

EDINBURGH, *continued*...... Kendall, W. B.
Kirk, J. B., M.D., *Bathgate*
Laycock, Thomas, M.D.
Lennox, D.
Library of University of Edinburgh
Livingstone, Bros.
Lucas, R., M.D., *Dalkeith*
M'Donald, A., M.D., F.R.C.S.E.
Macdonald, Keith, M.D.
Macgillivray, C. W., M.D.
Maclagan, D., M.D.
Maclaren, P. H., M.D.
M'Rae, E., M.B., *Penicuik*
Malcolm, R. B., M.D.
Middleton, J., M.D.
Muirhead, W. C., M.D.
Murdock, J., M.D.
Pattison, Thomas, M.D.
Playfair, J., M.D.
Royal College of Surgeons: per Dr. Inglis
Royal Medical Society of Edinburgh
Royal College of Physicians
Rutherford, W., M.D.
Rutherford, T., M.D., *Kelso*
Sanders, Prof., W. R.
Scott, Jas., *Bonnington*
Sibbald, J., M.D.
Sidey, J. A., M.D.
Silke, G. B., M.D.
Simpson, Prof. A., M.D.
Sinclair, Alexander J., M.D.
Stewart, T. G., M.D.
Stewart, D. E., *Selkirk*
Stewart, J., M.D.
Sym, Allan
Taylor, W., M.D.
Thomson, A., M.D.
Underhill, C. E., M.D.
Valentine, Rev. Colin S., LL.D., F.R.C.S.E.
Walker, J.
Watt, J. D., M.D.
Weir, T. G., M.D.
Woodhead, G. S.
Young, Peter, M.D.
Young, P. A., M.D., *Portobello*
Zeigler, W., M.D.

ELGIN .... DUFF, G., M.D., *Loc. Sec.*
George, J. T., M.D., *Keith, Banff*

Elgin, *continued*	M'Kay, Morris, M.D.
	Gallatly, W., M.D.
Forfar	Murray, W. F., M.D., *Loc. Sec.*
	Maclagan, Wedderburn, M.D.
	Hunter, C., M.D.
Fort William	Hutchinson, J., M.B.
Fraserburgh	Grieve, A. C.
Fyvie	Greig, A. F.
Galashiels	Somerville, R., M.D.
Glasgow	Anderson, J. W., M.D., *Loc. Sec.*
	Adams, J., M.D.
	Agnew, D., M.D.
	Anderson, T. M'Call, M.D.
	Brown, A., M.D.
	Buchanan, T. D., M.D.
	Buchanan, J. M., M.D.
	Burns, J., M.D.
	Carmichael, N., M.D.
	Carr, W., M.D.
	Cleland, J., M.D.
	Charteris, M., M.D.
	Coats, Joseph, M.D.
	Connell, R., M.D.
	Core, W., M.D.
	Cowan, R., M.D.
	Dick, J., M.D.
	Duncan, Eben., M.D.
	Farquharson, J., M.D., *Coatbridge*
	Fergus, A., M.D.
	Finlayson, J., M.D.
	Fleming, W. J., M.D.
	Forrest, R. W.
	Frew, —, M.D., *Newmilns*
	Fullerton, Neil, M.D., *Lamlash*
	Gairdner, W., M.D.
	Glasgow University Medico-Chirurgical Society
	Goff, B., M.D., *Bothwell*
	Gray, J., M.D.
	Henderson, T. B., M.D., 17 *Elmbank Crescent*
	Henderson, T. B., 48 *Kelvin Grove Street*
	Hunter, Walter, M.B.
	Lapraik, T., M.D.
	Lawrie, J., M.D.
	Leishman, W., M.D.
	Love, J. K.

GLASGOW, *continued* ...... Library of University of Glasgow
Library of Glasgow Faculty of Medicine
Library of Faculty of Physicians and Surgeons
M'Gavin, J., *Dennistown*
M'Fie, Johnstone, M.D.
M'Conville, J., M.D.
M'Colman, D., M.D., *Ballachulish*
M'Leod, G. H. B., M.D.
M'Vean, J. D.
Miller, J., M.D., *Springburn*
Millar, J., M.D., *Wishaw*
M'Millan, E., M.D., *Tradeston*
Morton, J., M.D.
M'Phail, D., M.D.
M'Kenzie, Henry
Muir, W., M.D., *Bridgeton*
Munro, Donald
Nairn, J. Stewart, M.D.
Napier, —, M.D., *Crosshill*
Newman, —, *Partick*
Patrick, W., M.D.
Peden, W. K.
Perry, R., M.D.
Pollock, A. B.
Pollok, R., M.D., *Pollokshields*
Reid, Thomas, M.D.
Reid, W. L., M.D.
Rigby, John
Ritchie, A., M.D., *Pollockshaws*
Robertson, A., M.D.
Robertson, A. M.
Semple, R., M.D.
Sewell, W. R., M.D., *Kincardine-on-Forth*
Simpson, P. A., M.D.
Smith, W., M.D.
Smith, Thomas
Spiers, D., M.D.
Steven, J. L.
Steven, Finlay, M.D., *Coatbridge*
Thomas, M., M.D.
Thomson, A. T., M.D.
Turner, H., M.D.
Walker, R.
Wallace, A., M.D.
Watson, W. R., *Dennistown*
Wilson, J. G., M.D.
Yeaman, G., M.D.

Glasgow, *continued*	Young, David, M.D., *Pathhead*
Glenluce	M'Cormack, W.
Greenock	Wallace, J., M.D., *Loc. Sec.*
	Black, J. R., M.D.
	Auld, Charles, M.D.
	Whiteford, —, M.D.
	Carlyle, J., M.D., *Auchmull*
	Marshall, W. J., M.D.
	M'Dougal, J., M.D.
	Paton, J., M.D.
	Douglas, J., M.D.
	M'Kecknie, W., M.D., *Thorndean*
	Stewart, A. D.
	Wilson, W. A., M.D.
Haddington..................	Howden, T., Jun., M.D., *Loc. Sec.*
	Martine, W., M.D.
Hamilton.....................	Loudon, J., M.D.
Irvine	Wilson, W., M.D.
Inverness	M'Nee, J., M.D.
	Aitken, Thomas
	M'Donald, W., M.B.
Jedburgh.....................	Blair, W., M.D.
Johnstone	Taylor, M. H., M.B.
Kelso	Johnson, E., M.D.
Kilcreggan..................	Zair, J. M., M.B.
Kirkcaldy	Gordon, H., M.D.
Kilmarnock..................	Macfarlane, A. W., M.D., *Loc. Sec.*
	Rankin, Guthrie
	Baxter, W., M.D.
Kilmaurs.....................	Buchan, George
Kirkcudbright	Urquhart, Andrew J.
Leith	Struthers, J., M.D., *Loc. Sec.*
	Henderson, J., M.D.
	Garland, O. H., M.D.
Leshmahagow	Lindsay, J.
Leuchars *see* St. Andrews	
Linlithgow..................	Hunter, G., M.D., *Loc. Sec.*
Lochgilphead...............	Cameron, J., M.D.
Melrose	Meikle, J., M.D.
Montrose	Howden, J. C., M.D., *Loc. Sec.*
	Steele, G., M.D.
Musselburgh	Scott, T. R., M.D.
Newburgh *see* St. Andrews	

New Galloway	Millman, A. M'Kinlay, M.D.
Newport *see* St. Andrews	
Paisley	Taylor, D., M.D., *Loc. Sec.*
	Donald, J. T., M.D.
	Fraser, Donald, M.D.
	Graham, Thomas, M.D.
	Infirmary Library
	M'Kinlay, —, M.D., *Barrhead*
	Paton, J., M.D.
	Richmond, D., M.D.
Penicuik	M'Rae, A. E., M.D.
Perth	Stirling, D. H., M.D., *Loc. Sec.*
	Bramwell, J. P., M.D.
	Irvine, W. S., M.D., *Pitlochry*
	Roy, W., M.D.
	Laing, H. W., M.B., *Bridge of Earn*
Peterhead	Jamieson, P.
Pitlochry *see* Perth	
Port William	
Rutherglen	Gorman, J.
St. Andrews	Mackie, J., M.D., *Loc. Sec.*
	Archibald, D., M.D.
	Library of University of St. Andrews
	Constable, J., M.D., *Leuchars*
	Niven, T., M.D., *Newburgh*
	Stewart, J., M.D., *Newport*
	Whitelaw, W., M.D.
Shotts	Caldwell, J.
Skelmorlie	Wylie, W., M.D.
Strathpeffer	Middleton, James, M.D.
Stirling	Gibson, C., M.D., *Loc. Sec.*
	Johnstone, W., M.D.
	Rae, J., M.D.
	Robertson, J., M.D., *Bannockburn*
Taynuilt	M'Nacnaughton, Allan
Tombeg	M'Rae, D., M.D.
Whithorn	Douglas, J. C., M.D.
Wick	
Wigton	M'Bride, Charles, M.D.

IRELAND.

ANTRIM	Adams, J. J., M.D.
ARDEE	
ARMAGH	Cuming, Thomas, M.D.
	Fraser, H., M.D.
	Huston, R. Todd, M.D., *Tyman*
	Palmer, J., M.D.
	Pratt, T., M.D.
ATHLONE	Langstaff, H. H., M.D.
AUGHNACLOY	Scott, W., M.D.
	Cordner, Louis, M., L.K.Q.C.P.
BAGNALSTOWN	Allen, C. D., M.B.
BALLINA	Macaulay, R., M.D.
BELFAST	Aiken, W., M.D.
	Cuming, Prof. D.
	Purdon, C. D., M.D.
	Purdon, T. H., M.D.
	Whitla, W., M.D.
	Workman, C., M.D.
	Hartree, J, P., M.D.
	Byers, J. W., M.A., M.D.
	Esler, R., M.D.
	Ulster Medical Society
BELLAGHY	Charles, D. Allen, M.D.
BOYLE	O'Farrell, H., M.D.
Bray *see* Dublin	
BRUFF	Macnamara, P. J., M.D.
CARLOW	O'Meara, Thomas P., M.D.
	O'Meara, W. H., M.D.
CARRICK-ON-SUIR	MARTIN, J., M.D., *Loc. Sec.*
	Reynett, J., M.D., *Portlaw*
	White, T. K., M.D., *Kilsheelen*
CASTLEISLAND	Nolan, W.
CASTLEWELLAN	
CAVAN	MALCOMSON, W., M.D., *Loc. Sec.*
COOKSTOWN	M'Iver, W., M.D.
CORK	FINN, E., M.D., *Loc. Sec.*
	Cremen, Patrick J., M.D.
	Donovan, Dennis, M.D.
	Harvey, J. R., M.D. (for Medical Club)
	Hobart, N., M.D.
	Jones, M'Naughton, M.D.
	O'Sullivan, S., M.D.
	Queen's College, Cork: per J. England, Esq.

CORK, *continued*	Tanner, W. K., M.D.
	Townsend, E. R., M.D.
	Townsend, W. C., M.D.
	Mulcahy, D.
COROFIN, Co. Clare	Macnamara, G. U.
CULDAFF	Gilmore, T. C., M.D.
DOWNPATRICK	Maconchy, J. K., M.B.
DUBLIN........................	MOORE, J. W., M.D., *Loc. Sec.*
	Armstrong, J. H.
	Banks, J. T., M.D.
	Barker, W. O., M.D.
	Barton, J. K., M.D.
	Beatty, J. G., M.D.
	Bennett, E. H., M.B.
	Benson, J. Hawtry, M.D.
	Biggar, S. L., M.D.
	Brady, J., M.D.
	Catholic University School of Medicine
	Churchill, F., M.D.
	Colles, W., M.B.
	Corley, A. H.
	Cruise, J. R., M.D.
	Cryan, R., M.D.
	Daly, F. H.
	Darby, T., M.D., *Bray*
	Davison, H. A.
	Denham, J., M.D.
	Duffey, G. F., M.D.
	Elliott, W. A.
	Fitzgibbon, H., M.D.
	Fitkgerald, C. E.
	Fitzpatrick, T., M.D.
	Foot, A. W., M.D.
	Grimshaw, T. W.
	Hamilton, E., M.B.
	Hayden, T., M.D.
	Head, H. H., M.D.
	Hudson, A., M.D.
	Jennings, W. B.
	Johnston, G., M.D.
	Kelly, J.
	Kidd, G. H., M.D.
	King's & Queen's College of Physicians
	Kirkpatrick, J. R., M.D.
	Library of Meath Hospital
	Library of Adelaide Hospital
	Library of School of Physic in Ireland
	Little, J., M.D.

DUBLIN, *continued*	Madden, T. W., M.D.
	Marks, A. H., M.D.
	Mason, Thomas
	M'Dowell, B. G., M.D.
	M'Donnell, J., M.D.
	Minchin, H., M.D.
	Moore, C. F., M.D.
	Moore, W., M.D.
	Murphy, J.
	National Library of Ireland
	Powell, G. W., M.D.
	Pollock, J. F., *Blackrock*
	Peele, E., M.D.
	Porter, G. H., M.D.
	Purefoy, R. D., M.D.
	Rainsford, R., M.D.
	Richmond Hospital Library
	Royal College of Surgeons
	Shannon, P., M.D.
	Smyley, P. C.
	Stoker, W. T., M.D.
	Thomson, W., M.D.
	Torney, Thomas, M.D.
	Tottrell, W.
	Trinity College Library
	Wharton, J. H., M.B.
	Wheeler, W. J., M.D.
DUNDALK	
ENNISCORTHY	Drapes, Thomas, M.D.
ENNISKERRY..................	Barrington, C. E., M.D.
GALWAY	Queen's College
	Kinkead, Prof., M.D.
GLASSLOUGH..................	Stewart, R. W., M.D.
KILKENNY......................	Johnson, Z., M.D.
KILLALA	
KINVARA	Nally, W. J.
KINGSTOWN	ADAMS, W. O'BRIEN, *Loc. Sec.*
LIFFORD	Little, Robert, M.B.
LIMERICK	KANE, THOMAS, M.D., *Loc. Sec.*
	De Landre, G., M.B.
	Courtenay, E. Maziere
LONDONDERRY	BERNARD, WALTER, M.D., *Loc. Sec.*
	Miller, J. E., M.D.
	White, Barnwall, M.D.

Londonderry, *continued*...	M'Cullagh, Jas. A., M.D.
Loughrea	
Maghera	M'Gowan, W., M.D.
Malahide	Lloyd, Hans, M.D.
Monaghan	Ross, D. M., *Loc. Sec.*
	Robertson, J. C., M.D.
Moynalty	Dundas, G. H.
Navan	Hamilton, F., M.D.
Newcastle	Clarendon, S., M.B.
Oranmore	Geoghegan, Lawrence
Pomeroy	Henry, R., M.D.
Roscommon	Harrison, J., M.D., *Loc. Sec.*
Roscrea	Powell, B. C., M.D., *Parkmoor*
Rosstrevor	Vesey, T. A., M.B., *Loc. Sec.*
Rostellan	Travers, R. B., F.R.C.S.
Sligo	M'Munn, J., M.D.
Tullow	Warren, W., M.D.
	Kidd, H.
Wexford	Boxwell, H. H., M.D., *Loc. Sec.*
	Cardiff, J. R., M.D., *Ballinabola*

EUROPE.

Bologna (Italy)	Ciaccio, Guiseppe, M.D.
Como (Switzerland)	Comolli, —, M.D.
Constantinople (Turkey)	Patterson, John
Florence (Italy)...........	Young, D., M.D., *Loc. Sec.*
Kazan (Russia)	Inyasevsky, N. J., M.D.
Madeira	Grabham, T., M.D.
Marseilles (France)	Duranty, E. Nicolas, M.D.
Nice (France)..............	Crossby, —, M.D.
Paris	Pozzi, S., M.D., *Loc. Sec.*
Vienna....................	Roxburgh, R., M.D.

AFRICA.

Capetown	Gorman, Charles, M.D.
East London (Cape Col.)	Hartley, W. Darley
Graaf Reinett (Cape Col.)	Hislop, G. B., M.D.
Port Elizabeth (Algoa Bay)	Neebe, C. W., M.D.
Somerset East (S. Africa)	Clarke, Thomas Furze, M.D.
Winburg (Orange Free State)	Leech, J. R., M.D.

ASIA.

Bombay	Banat, H. E.
	Dalal, K. A., M.D.
	Joynt, F. G.
	Joynt, C., M.D.
	And others
Belgaum (Bombay)	Wall, Robert M.
Bareilly	Loch, J. H., M.D.
Calcutta	Jones, J., M.D.
Ceylon (Colombo)	Kinsey, W. R., M.D.
Hong-Kong	General Military Hospital
Madras	Brockman, Surgeon-Major E. F., *L. Sec.*
	King, H., M.B.
	Morgan, W. H.
	Lloyd, E. E.
	Porter, A., M.D., F.R.C.S.I.
	Robertson, C., M.D.
	Cornish, W. R., F.R.C.S.
	Sibthorpe, C.
	Branfoot, A. M., M.B.
	Nath, J.
	Sturmer, A. J.
	Moran, J. J., M.D.
	Ward, T.
	Boon, H.
	Andy, S. Pulney, M.D.
	Mootoosomar, Moodehar
Moultan	Blood, Joseph, M.B., *Loc. Sec.*

MOULTAN, *continued*	Mullane, J., M.D.
	Juswant, Rai
MYSORE	M'Gann, T. J.
NAGPUR	School of Medicine
NANII TAL (N. W. P.) ...	Cleghorn, James, M.D.
ORISSA	Gupta, B.
PESHAWAR	Haig, P. E. D. H.
SINGAPORE	Rowell, T. S., M.D.
VIZAGAPATAM	Smith, J.
YOKOHAMA & YEDO (Japan)	ELDRIDGE, STUART, M.D., *Loc. Sec.*

SOUTH AUSTRALIA.

ADELAIDE	Thomas, J. D., M.D.
	Whittell, Horatio, M.D.
	Way, E. Willis, M.D.
ARMIDALE....................	Wigan, George, M.D.
ARARAT....................	Morrison, A.

VICTORIA.

MELBOURNE	BARKER, E., M.D., *Loc. Sec.*
	And nineteen Members

NEW SOUTH WALES.

SYDNEY	SPARK, JAS., M.D., *Loc. Sec.*
	Booker, R. R., M.D.
	Dixson, Craig, M.D.
	Durham, J. C. C., M.D.
	Pickburn, —, M.D.
	Mackenzie, W. F., M.D.
	M'Kellar, Charles, M.D.

NEW ZEALAND.

AUCKLAND	Auckland Institute
CANTERBURY	Guthrie, T. Orr, M.D.

CHRISTCHURCH	IRVING, J., M.D., *Loc. Sec.*
	Prins, —.
	Townend, J. H., L.R.C.P.
NAPIER	DE LISLE, F. L., M.D., *Loc. Sec.*
	Caro, —, M.D.
	Hitchings, Thomas, M.D.
	Spenser, W. J.
	Todd, Alex., M.D., *Waipawa*
	Menzies, —, M.D.
NELSON	Squire, W. W., M.D.
	Williams, G.

QUEENSLAND.

BRISBANE	FLOOD, S., M.D., *Loc. Sec.*

TASMANIA.

SWANSEA	Lovett, —, M.D.

BARBADOES.

	WALLCOTT, R., M.D., *Loc. Sec.*
	Garrison Library, St. Ann's

UNITED STATES.

ABINGDON.....................	REECE, MADISON, M.D., *Loc. Sec.*
BOSTON	SALTER, W. H., M.D., *Loc. Sec.*
	And thirty-three Members
BUFFALO	Cronyn, J., M.D.
	Samo, J. B., M.D.
DUBUGUE	Horr, Asa, M.D.
	Hay, Walter, M.D.
JERSEY CITY	Watson, B. A., M.D.
LOUISVILLE (Kentucky) ...	Library of the Polytechnic Society of Kentucky
LYNN (Massachussetts) ...	Pinkham, J. G., M.D.
SING-SING	Fisher, G. J., M.D.
	Fisher, A. K.
	Helm, W. A., M.D.

New York	Wood & Co. (Messrs.), *Local Agents*
	Knight, C. H., M.D.
	And forty Members
Philadelphia	Blackiston, Presley, *Local Agent*
	Da Costa, J. M., *Pennsylvania*
	Meigs, John Forsyth, *Pennsylvania*
	Lewis, Samuel, *Pennsylvania*
	Hay, Thomas, *Pennsylvania*
	Ashurst, John, jun., *Pennsylvania*
	Stillé, Alfred, *Pennsylvania*
	Collins, James, *Pennsylvania*
	Atlee, W. L., *Pennsylvania*
	West, T. H., *West Virginia*
	Kemper, G. H., *Indiana*
	Otte, C. H., *Maryland*
	Moses, T. A., *Missouri*
	Benham, S. M., *Pennsylvania*
	Schultze, S. S., *Pennsylvania*
	Curwen, John, *Pennsylvania*
	Darrach, James, *Pennsylvania*
	Palmer, J. Dabrey, *Florida*
	Gaines, E. P., *Alabama*
	Fleming, A., *Pennsylvania*
	Prince, David, *Illinois*
	Hewlett, V. P., *New Jersey*
	Hearne, J. C., *Missouri*
	Wey, W. C., *New York*
	Gamble, D. C., *Missouri*
	Van der Veer, A., *New York*
	Carpenter, John S., *Pennsylvania*
	Surgeon-General, *U. S. Navy*
	State Hospital for Insane, *Warren, Pennsylvania*
	Van der Poel, S. O., *New York*

CANADA.

Durham	Gun, Jas., M.D.
Montreal	Nichol, Thos., M.D., LL.B., B.C.L.
Nova Scotia	Maclarty, —, *Sydney*
St. Johns (N. Brunswick)	Berryman, —, M.D.
Toronto	Geikie, W. B., M.D.

WEST INDIES.

St. Lucia .... Galgey, Otho, M.D.

SOUTH AMERICA.

Valparaiso .... Cooper, G. F., M.D.

Argentine Republic ...... Frend, J. A., *Rosario*

JAMAICA.

Kingston .... Anderson, J., M.D.
Wegg, John A.
Clarke, J. H., M.D.
Saunders, A. R., M.D.

www.ingramcontent.com/pod-product-compliance
Lightning Source LLC
Chambersburg PA
CBHW020909230426
43666CB00008B/1376
* 9 7 8 3 7 4 4 6 7 0 8 6 9 *